What is edible?

What is edible?

A journey through the world of food taboos and food cultures

By

Adel P. den Hartog

Edited by

Annemarie de Knecht-van Eekelen and Jon Verriet

BRILL | WAGENINGEN ACADEMIC

Cover illustration: Vegetable market, Castle Groeneveld in Baarn, The Netherlands. Photo by M. Sappok.

The Library of Congress Cataloging-in-Publication Data is available online at https://catalog.loc.gov
LC record available at https://lccn.loc.gov/2024029216

Typeface for the Latin, Greek, and Cyrillic scripts: "Brill". See and download: brill.com/brill-typeface.

ISBN 978-90-04-69431-6 (paperback)
ISBN 978-90-04-70157-1 (e-book)
DOI 10.3920/9789004701571

Contents

Accountability

In the 1980s, Annemarie de Knecht-van Eekelen worked on her PhD thesis on the history of infant formula, while Adel den Hartog wrote his dissertation on milk as a food product in Indonesia. Their shared interest in food history led to contact, which intensified when they started participating in the meetings of the International Commission for Research into European Food History (ICREFH). This society had been founded in 1989 and consisted of a small group of European scholars. The lectures delivered at the annual meetings were compiled in a series of books, of which Adel edited the volume *Food technology, science and marketing: European diet in the twentieth century* (1995).

A few years after Adel's death, Annemarie met his widow, Frida den Hartog-Koet. She told her about a bequeathed manuscript by Adel that he had still wanted to publish. She thought it would be a good idea for Annemarie to take a look at it. But as these things go, it stayed with Annemarie until she came into contact with Jon Verriet in early 2021. He had published an article on the history of the Food Information Agency with ample coverage of Adel's father's actions, and years before had once spoken to Adel himself.

Together, we took up the manuscript of Adel. We maintained the original text, which had been completed in 2010–2011, as much as possible. We elaborated on some themes in separate boxes and added recent data here and there. Furthermore, we have provided the book with illustrations. At the back of the book is the literature Adel consulted. References we added are in the footnotes. With a short epilogue, we have given Adel den Hartog's work a fitting conclusion.

Our thanks go to the Van Dam Foundation for making this publication possible. Dr Frida den Hartog-Koet has helped us greatly by making books and photo material available. Many thanks for that. Cooperation with the team of Wageningen Academic Publishers was particularly pleasant. The appearance of the book is entirely their responsibility. We hope the book will appeal to the general culturally interested reader. In addition, it can be used as a handbook for those who want to understand the background of our eating habits. This knowledge can be used in a range of nutrition and lifestyle interventions.

Annemarie de Knecht-van Eekelen and Jon Verriet
Malden, September 2021

Word in advance

Sometime in the spring of 2000, Adel and I were sitting at a beautifully laid table in the professors' restaurant of the Free University of Brussels. The aim was to smoothly hand over the chairmanship of the International Committee for Research into European Food History (ICREFH). ICREFH's aim was research into the history of food in Europe from the late 18th century onwards. Adel was chairman from 1995 to 1999; I succeeded him. Organising a biennial colloquium and the related book were the main concern of ICREFH's president, that much I knew. But I wanted to know more about the practical side (Where does the money come from? Who decides on the central theme? How is the selection of participants done?) and, above all, what about the relationships between renowned researchers like John Burnett, Hans-Jürgen Teuteberg, Derek Oddy, Alexander Fenton, Eszter Kisbàn or Roman Sandgruber, all members of the club's first hour. On the practical questions, Adel enlightened me quickly and effectively, but my second concern required a bit more time (and emptying the bottle of wine). Indeed, relations among ICREFH seniors were not always smooth, which Adel tactfully explained. That meal taught me a lot not only about organising international colloquia but also about interpersonal relations. And above all, I got to know Adel as a particularly erudite, sensitive and empathetic intellectual with a great talent for languages.

Talking to Adel was always instructive. I remember a conversation during a long bus ride in the Czech Republic (ICREFH colloquium, 2003). Adel and I talked about the place of food history in general history and in *food studies*, where multidisciplinarity was a central issue. Adel, as a geographer and nutritionist, argued for true interdisciplinarity while I, a social historian, emphasised one's own discipline, however fertilised with other disciplines. We were both right, we decided.

Münster 1989, London 1991, Wageningen 1993, Vevey 1995, Aberdeen 1997, Tampere 1999, Alden Biesen (Limburg) 2001, Prague 2003, Berlin 2005, Paris 2009, Bologna 2011: ICREFH colloquia where Adel presented papers, chaired sessions, chaired congresses, debated, concluded, conversed and displayed his talent for nuance and humour. The variety of his contributions was immense and illustrated his interdisciplinary approach. He talked about schoolchildren's meals, the place of dietetics in advertisements, the modern packaging of milk, the importance of vegetables, healthy biscuits, the technology of eating out, the discovery of the vitamin or domestic education. In each case, these were topics that were relatively new, pre-eminently in the Low Countries.

Although the book you hold in your hands was written 10 years ago, it is still relevant today because it offers a synthesis of the history of nutrition in the Netherlands, has a clear and global question (what does a society consider edible?), is the result of years of historical research by an exact scientist, and addresses current themes (eating differently, for example). The book differs from classics on Dutch food history, brought by the likes of Anneke van Otterloo, Jozien Jobse-van Putten, Ineke Strouken, Paul Spapens, and Adel himself (*De voeding van Nederland in de twintigste eeuw* from 2001), but will undoubtedly soon become as *incontournable* as the aforementioned classics.

Em. Prof. Peter Scholliers
Brussels, September 2021

Introduction

Here's an experience that often occurs while travelling, especially to far-away countries. You are served a meal, and you wonder, 'What is this, is it edible?' True culinary adventurers will try to eat anything foreign and unfamiliar, but most of us prefer to take a look at it first. Sure, humans are omnivores, but they do not eat everything available in terms of food. Globally, people have indeed eaten everything at one time or another and have continued to eat these things if they did not make them immediately sick or kill them. But this is not the case at a regional or local level.

My curiosity about why a certain food is valued in one culture and loathed in another arose during my work at the FAO, the Food and Agriculture Organization of the United Nations, first in Africa and then at its headquarters in Rome. In Wageningen, at the Department of Human Nutrition at Wageningen University, I also got to know other food cultures. International teaching at the university, my research in Africa and Asia, and my contacts with colleagues and students allowed me to address the diversity of ideas about food in diverse cultures. Wageningen students encouraged me to think about why we eat differently in the Netherlands and Europe than in non-Western countries and how mutual influence takes place.

Human history has seen an interesting shift. Over time, people started eating not just to stay alive, but they increasingly lived to enjoy food. Food has taken on great cultural significance. Eating communally can foster communication with others or emphasise the identity of a group or society. French anthropologist Claude Lévi-Strauss (1908–2009) said that food should be good to eat, but above all it should be good enough to make us think.

American psychology professor Paul Rozin (1936-) introduced 'the paradox of the omnivore', by which he meant that, as omnivores, humans can eat anything, but still have trouble making food choices. There are two opposing feelings: on the one hand, there is the need to try something new; on the other, there is the fear of the unknown. To survive in a changing environment, it is necessary to try unknown foods and overcome the fear and repulsion. In the Netherlands, we say literally 'what the farmer doesn't know he doesn't eat' (the equivalent in English would be 'one man's meat is another man's poison'). Yet over time, even that farmer has learned to eat other kinds of food.

The central theme of this book is the question of what is edible and non-edible in a society. Why is something edible to us and not edible in other areas and cultures? What are the underlying processes that influence the concept of 'edible'? The diversity of food and eating cultures is great, so limits are inevitable. I will leave aside individual likes and dislikes.

The first chapter of this book provides a geographical exploration of why we eat what we eat. Then, in Chapter 2, I look at the phenomenon of edible and inedible on the basis of the spread of food over time, from one continent to another—I call this food diffusion. This chapter looks at the acceptance of new food products in other food cultures, globalisation, local reactions and reluctant acceptance, and the long road from unknown to 'good to eat'.

In Chapter 3 I discuss the phenomenon of food taboos, especially in the form of food that is unclean or too holy to eat. Chapter 4 is then devoted to the eating of meat, the most controversial food. In the next chapter on insects, rats and dogs, I explain that disgust associated with certain foods is culturally determined. For the Dutch, milk was the white engine, but that is certainly not true in other parts of the world—that is what Chapter 6 is about.

CHAPTER 1

Why do we eat what we eat?

Unknown and unloved

Our food choices have a physiological basis. Hunger and appetite are the major regulators of these choices that are controlled by the central nervous system. After all, the main function of food is to sustain the body. Food is also eaten because it is appetising or can provide status. But 'food' is a relative term (Table 1). In the last 10,000 years, no new edible plants and animals have been discovered that have not already been eaten somewhere in the world. But locally, people by no means eat all those plants and animals. What is considered inedible in the Netherlands will be found on a plate in other countries and cultures. What is considered edible, tasty or status-enhancing varies widely between different countries and cultures.

There are two basic reasons why an animal or plant is or is not seen as edible in a society. First, the food may be subject to a ban or taboo. I write about this in Chapter 3. The second reason is that the food is wholly or largely unknown. Unfamiliarity with food causes timidity or even suspicion. Sometimes a food was totally unknown, like the potato in Europe, before Columbus 'discovered' America. The plant was put on display as a botanical speciality in the Hortus botanical gardens of Leiden University. Now it is impossible to imagine our food culture without the potato; unknown eventually became loved.

Examples of foods that were known at some level are spices and sugar. In the late Middle Ages, these were used in the kitchens of Dutch nobility and wealthy citizens, but not by other population groups. Ignorance can also be regional: nomadic cattle breeders in Africa do not eat fish because fish is not found in their environment.

TABLE 1   Food is a relative concept: some examples

Animal/product	Generally edible in (for example)	Generally inedible in (for example)
dog	West Africa, Philippines, China	Netherlands
horse	Vietnam, Netherlands, Belgium, France	UK
rat	tropical Africa, Thailand	Netherlands
insects: caterpillar, termite, maggot	Africa, Southeast Asia, Mexico, South America	Netherlands
cow	Netherlands, Europe, America	India (among Hindus)

© ADEL P. DEN HARTOG, 2024 | DOI:10.3920/9789004701571_003

People are usually willing to eat an unfamiliar dish as long as they don't know what the ingredients are. This can change if people find out that the sauce they are eating is made of insects or the meat is that of a dog, for example. Psychosomatic reactions then start to play a role: the diner gets a general feeling of uneasiness and sometimes even nausea. Interestingly, eating unfamiliar plant foods, such as vegetables and fruit, triggers far fewer emotional reactions than eating unfamiliar meat. A reason to take a closer look at humans as carnivores.

Humans as carnivores

Humans can eat both animals and plants, making them flexible compared to animals that depend on a particular plant or animal species. An extreme example is the giant panda, which only eats bamboo. When the bamboo forest is depleted, the giant panda dies of starvation. Because humans are omnivorous they have over time been able to settle anywhere in the world, provided there are enough edible animals and plants.

FIGURE 1   The Bull (1647), painting by Paulus Potter (1625–1654). The very large size of the painting (235.5 × 339 cm) offers a 'glorification of cattle' of the kind only seen in Holland.
SOURCE: WIKIPEDIA. THE WORK IS PART OF THE MAURITSHUIS COLLECTION, WIKI COMMONS

The place of meat in people's diets is largely determined by geographic factors, i.e. the extent to which edible animals are available. In the long history of human development, starting with hunter-gatherers, animal products have always been part of the diet, but the quantities available were closely related to circumstances.

Compared to their neighbours, the average Dutchman is not a big meat eater. Perhaps this has to do with the special position of dairy in the Netherlands. The provinces of Holland and Utrecht were important production areas for cheese, butter and milk at the end of the Middle Ages. This was linked to the soil structure: the wet fields were not suitable for growing grains, so farmers were forced to turn to animal husbandry (Figure 1). This led to a boom in dairy exports and the simultaneous import of cereals for food supply. Dairy gained an important place in Dutch food culture at that time.

In French, German and Austrian cuisine, meat has a central position in the hot meal. The high regard for meat in France can be illustrated by the changing meaning of the French word for meat, *viande*. In the Middle Ages, the French word for meat was *chair* (flesh). The word *viande*, derived from the Latin *vivenda* ('everything necessary for life'), referred more generally to foodstuffs, not only meat, but also dairy, vegetables, fruit and grains. Over time, the word *viande*, food necessary for life, referred only to meat.

Meat and food science

The importance of meat in the diet took on a scientific dimension in the nineteenth century. By combining natural science principles with new technologies, researchers began to understand the (bio)chemical and physiological significance of food for human health. Meat was then seen as the main source of protein—not surprisingly, as malnutrition was a major problem in nineteenth-century Europe.

Meat, as a provider of essential dietary protein, remained the basis of a decent hot meal well into the 20th century. German physiologists Carl von Voit (1831–1908) and his pupil Max Rubner (1854–1932) determined that an adult working man needed about 145 to 159 grams of protein per day. Indeed, it had long been assumed, incorrectly, that protein plays a role in metabolism in muscle work. Based on this, the ration of a German soldier in 1914 consisted of 375 grams of meat per day. In our view, that is a huge and completely unnecessary amount. As a matter of fact, there was also an ideological motive for the German war ration of 1914. The people back home had to be convinced that the men at the front were being well taken care of, and lots of meat was a symbol of this care. In the end, the war lasted much longer than expected. For Germany, the great fixation on meat therefore meant that a lot of agricultural land had to be used for meat production, at the expense of growing potatoes and cereals for the civilian population. In his article on the food supply of the German army, Dr Peter Lummel, director of the Domäne Dahlem open-air museum in Berlin, cites this as one of the reasons for the serious food shortages at the end of World War I in Germany.

Humans as man-eaters

When humans eat other humans, it is called cannibalism or anthropophagy. Stories about man-eaters are popular in legends and fairy tales. However, the phenomenon is so extremely rare that I won't go too deeply into it. Most people are horrified by the idea of eating human flesh. The term 'cannibalism' comes from the Caribbean. The people of the West Indies known as the Caribs were regarded by the Spanish as savage, bloodthirsty man-eaters. In reality, human flesh from vanquished opponents was sometimes eaten during ritual customs, but not as a result of gluttony and other fantastical customs. This perception was partly what led to the Spaniards subjugating and eventually exterminating these peoples.

In the widely read travelogue of the German soldier and sailor Hans Staden (*c.*1525–1579), the latter recounts how he was captured by the Tupinambá, an Indian tribe in Brazil, which, according to Staden, practised cannibalism (Figure 2). His book, which was published in an English adaptation in 2008, contributed to Europe's perception of the New World. Cannibalism became part of that.

FIGURE 2   Cannibalism in Brazil (1557) according to Hans Staden, who is—stripped of his clothes—in the background. Engraving by Théodore de Bry (1528–1598) from the year 1562
WIKI COMMONS

The comic book *Tintin in Africa*, the first edition of which appeared in Dutch in 1946 under the title *Tintin in Congo,* contains the clichéd image of Europeans in a big black cooking pot. It is an example of condescending ethnocentrism towards sub-Saharan Africa. American anthropologist William Arens (1940–2019), in his in-depth study of the phenomenon of cannibalism, concluded that abhorrence of eating human flesh is universal. This also applies to the areas known in (Western) imagery for eating human flesh: sixteenth-century Aztec Mexico, present-day Africa and the cultures of New Guinea and Oceania.

Eating the dead in times of great famine is another matter. It happened in Europe in the early fourteenth century. The climate got colder, the so-called Little Ice Age arrived, and the beginning of this period was accompanied by crop failures and famine. The years 1315–1321 were marked by great mortality. From England, the Baltic region and Poland there were reports of people eating fellow humans who had died of starvation.

In American history, the fate of the Donner company is well known. A large group of people led by George Donner made their way through the mountains of Utah in the winter of 1846–1847. Men, women and children were on their way to California in 23 vehicles. The group was beset by a blizzard just before the peaks of the Sierra Nevada. When the rescue teams finally arrived, they were horrified to see signs of cannibalism.

A well-known example from the 20th century is the famine in Ukraine in the years 1932–1933, the Holodomor, 'death by hunger'. Millions of people died during that period; holocaust expert Timothy Snyder cites a figure of 7.5 million in his book *Bloodlands* (2010). The famine was the result of Stalin-enforced collectivisation of agriculture and excessive compulsory grain deliveries by Ukrainian farmers to the Soviet Union. The British journalist Gareth Jones (1905–1935) visited the famine zone in the 1930s and publicised the prevalence of cannibalism. However, many foreign correspondents in Moscow refused to believe Jones: it could not be true that such a thing was happening in a state under construction. Only much later did they acknowledge that Jones was right. His diaries were exhibited at the Cambridge library in 2009. Jones paid for his work with his life. He was murdered, probably by the Soviet secret service.

A final example of cannibalism in an emergency situation occurred after Uruguayan Air Force Flight 571 crashed in the Andes in 1972. The disaster was made famous by the numerous book and film adaptations of the survivors' stories. After the crash, the 45 surviving occupants had to sustain themselves for 72 days. In the end, 16 people survived by eating their deceased travelling companions. So, apart from a few pathological cases where killers ate a piece of their victim, anthropophagy occurs mainly in crisis situations.

Incidentally, the discussion about humans as eaters of their own kind was rekindled by archaeological finds from 2009. Prehistoric signs of cannibalism were found in a settlement of the Linear Pottery Culture from around 4950 BC near Herxheim in Germany. According to archaeologists, this was not about hunger or burial rituals, but rather war cannibalism, i.e. the slaughter and eating of prisoners. Over a period of about 50 years, about a thousand people were slaughtered and probably eaten. Nevertheless, eating human flesh remains historically an extremely marginal phenomenon that disgusts people in all cultures.

Inedible becomes edible

Humans have a great advantage over animals: not only are we omnivorous, but we can also cook our food. Dependence on raw food limits the number of food sources. When humans started using fire to heat food, a large number of indigestible plants became edible. This allowed food sources to expand exponentially. Without this technique, grains such as wheat, barley, rice and maize, for example, would not have become staple foods in large parts of the world. Another important advantage of heating is that dangerous forms of contamination by microorganisms could be avoided. Given the often unhygienic conditions, eating raw meat was not recommended in the past. In addition, several foods naturally contain toxins that are rendered harmless by heating. An example is the harmful *oxalic acid* present in rhubarb, sorrel, purslane and chard. After eating insufficiently heated legumes, symptoms of poisoning can occur, as beans contain the toxic substance *fasin* when raw. Another example is cassava, which in its raw state contains the toxic *hydrogen cyanide*.

Fire

Scientists attach great importance to the use of fire in food preparation. According to anthropologist Claude Lévi-Strauss, using fire and preparing food was the most important step in man's development in distinguishing himself from nature. An English anthropologist and primatologist, Richard Wrangham, elaborated on this idea in his book on cooking and the origins of man. He believes that cooking was the single most important leap forward in human evolution and that active fire control by humans is one of the essential characteristics that distinguish humans from primates and other highly developed mammals. His experiments with monkeys, by the way, show that when given a choice between raw and heated food, they also choose heated food that is softer and faster to digest.

According to Greek mythology, Prometheus stole fire from the Olympian gods and gave it to humans (Figure 3). This is not how it would have happened, but it remains a point of contention when people first started to use fire for preparing food. About 250,000 years ago, a new type of human appeared, the Neanderthal, very much related to modern Homo sapiens. Neanderthals largely depended on hunting for obtaining food. The genus Homo, the highest evolved humanoids, had until then settled in frost-free areas, but the Neanderthals extended their territory beyond the

FIGURE 3
Prometheus carries the fire (circa 1637)
PAINTING BY JAN COSSIERS (1600–1671),
PART OF THE COLLECTION OF THE PRADO IN
MADRID, WIKI COMMONS

northern boundary of the frost-free zone. Neanderthals often, but not always, used fire, probably also for preparing food. Only after the appearance of Homo sapiens, around 200,000 years ago, did making fire and heating food become a standard part of human civilisation. Around 40,000 years ago, mastery of fire became universal and food preparation by heating became a ubiquitous human activity.

Heating food has been instrumental in the evolution of our digestive system. It is generally believed that by eating cooked foods with a higher energy content and lower cellulose content, humanoids acquired a shorter intestinal system that is less suitable for digesting raw food.

Lack of fuel

In areas in developing countries where there is insufficient fuel, all kinds of problems arise around nutrition. Often, firewood is collected by women in fields and forests. It is usually too expensive to buy. In the absence of firewood, food preparation methods are changed. In our research in Malawi, for which Inge Brouwer received her PhD in 1994, we saw that women stopped cooking beans because the cooking time was too long and therefore required too much firewood. Omitting beans from the diet reduced the already low protein supply. Meals became simpler and simpler, and the number of hot meals per day was reduced from three to two and eventually to one. This was perceived as very unpleasant by the affected households. Incidentally, the study shows that there was no increase in the amount of raw food eaten as compensation in these cases.

Cooking methods

The development of cooking technology involves two different basic methods: the direct cooking method, whereby food comes into direct contact with the fire; and the indirect method. The direct method is the roasting of food, probably the oldest way of preparing food. It is still a popular cooking method, as in *kebabs* in the Middle East and *satay* in Indonesia.

The indirect method involves removing the food from direct contact with the fire. Throughout human history, very ingenious methods of preparation have been developed. For instance, a very old custom is to wrap the food in relatively fire-resistant, fresh leaves and cook them until tender in the hot ashes. In societies where pottery had not been developed, food was also stewed in carefully covered pits (called pit ovens) lined with glowing hot stones. Incidentally, this method is still used on traditional occasions on several islands in Oceania and by tribes in Brazil. Another method, developed in the Solomon Islands, involves filling wooden barrels with water which is then kept on the boil with hot stones.

In the Middle East, pottery technology was developed around 6500 BCE, at the beginning of the New Stone Age. This was an important step for food preparation. In earthenware pots, food could be prepared with water or oil without too much contamination from earth, ash or smoke. The food not only became cleaner but also changed in taste.

In a baking oven, hot air could also be used to bake bread or other products, such as *roti* in India. Baked bread became a supplement to, or substitute for, flour patties cooked with water. In Egypt, ovens were already in use around 3000 BCE.

What determines food choices?

What a population group does or does not eat and drink is mainly determined by four factors: geography, food security or sufficient access to food, food culture, and finally socioeconomic situation.

Geography

The geographical factor determines the way humans have arranged the environment for the production, processing and distribution of food. Climate, soil, altitude, countryside or city affect food availability and variation. In countries with distinct seasons, including a cold or dry season, and weak agriculture, food availability can fluctuate wildly. In some parts of West Africa, for example, during dry periods, food supplies run out at some point. Then the famine season breaks out. Women are forced to go into the field to collect 'famine food': seeds, tubers and fruits that they would otherwise ignore. Famine food is difficult to collect, difficult to prepare and unpleasant to eat.

In the age of air transport with built-in refrigerated containers, the geographical factor has become less important. Yet it has had a lasting influence on the food choices of several societies, including the Dutch. In our regions, the temperate climate determined what became our staple food. It allowed the potato, originating from the Andes, to become the staple food in north-western Europe (see Chapter 2). Cattle farming also does well in our temperate climate, and the dairy produced—milk and cheese—is an important component of the diet. In the humid tropics, where dairy farming is rarely possible, dairy is still absent from the daily diet. And although Dutch people in the colonies did become familiar with rice, it was eaten only rarely in Europe because it was difficult to grow there. These historical differences in geography have led to differences in food choices that (partly) persist today.

Food security

Food security means having year-round access to enough food to enable healthy living and working. Both food production and purchasing power are very decisive for food security. After all, food can be available without everyone having access to it. For instance, meat is highly valued in tropical Africa, but large parts of the population are vegetarian due to poverty. India has an ancient dairy culture, but the part of the population living below the poverty line knows dairy only by hearsay. In contrast, production opportunities and purchasing power have made Americans big meat eaters. However, the US is also a land of great contradictions where there are *food deserts*, areas where affordable and healthy food is limited. The term *food desert* stems from a 1995 English study of eating behaviour among low-income residents with few opportunities to obtain fresh, healthy food. The US Department of Agriculture *tracks* the number of *food deserts*. The map in Figure 4 shows relatively poor areas where many people have to travel 1 to 20 miles (1.6 to 32 km) for fresh food.

At Wageningen Economic Research, much research on food security has been conducted under the leadership of Professor Ruerd Ruben. In his view, it rests on three pillars: the nutritional value of food, the efficiency with which the food produced reaches people, and influencing healthy behaviour. Nutritional value can be increased by enriching food with (micro)nutrients. Efficient production is an issue of chain organisation. Frequently, progress is made in primary production at the beginning of the chain—higher yields, better quality—which is then undone further down the chain. Finally, Ruben thinks incentives to influence behaviour are desirable: in favour of the availability of high-quality food and in favour of more conscious and less wasteful consumption.

Food culture

Food culture, or eating habits and customs, greatly influences what is and is not eaten, with whom and why. In the Netherlands, a meal is characterised as 'simple

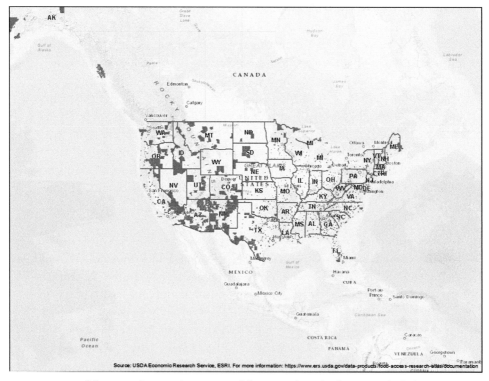

FIGURE 4   Food deserts in the United States. In red the areas where the distance to fresh food for many
             people is at least 1–20 miles.
             ECONOMIC RESEARCH SERVICE, U.S. DEPARTMENT OF AGRICULTURE, FOOD ACCESS
             RESEARCH ATLAS

but nutritious', a saying of cartoon character Olivier B. Bommel.[1] Jozien Jobse-van
Putten uses this phrase as the title for her PhD thesis on the history of Dutch food
habits. Food culture did change in the final decades of the 20th century. Simple
meals based on traditional Dutch dishes were increasingly being supplemented by
dishes from other countries in Europe such as Italy, Spain and Greece. Even more
recent are the influences from Turkey, North Africa, the Middle East and India.

An important part of food culture is determined by religions that stipulate
precepts and food laws about what can and cannot be eaten (see Chapter 3).
Interestingly, countries with a highly developed cuisine, such as France, Italy and
China, had a pronounced court culture in the past. The court had proportionally
unlimited access to a lot and a variety of food. Sophisticated dishes and court rules

---

1   In Marten Toonder's comic strip 'Tom Puss and the evil eye' (1961), Mr Bommel says: 'Prepare me
    a simple yet nutritious meal, so that I can go to work strengthened.' The phrase recurs regularly
    thereafter.

on food functioned as examples for the emerging bourgeoisie, thus gaining a foothold in food culture.

What is eaten or not eaten in a society is not static. On the contrary, it is constantly changing. The dynamics of this process are best illustrated by food diffusion, the spread of food from one continent to another. I will discuss this concept in the next chapter.

Socioeconomic situation

Food is one of our basic needs. First and foremost, humans, like animals, eat to stay alive. Millions of people around the world do not have enough food. The FAO estimated for the period 1995–1997 that 826 million people on earth were underfed. Most hunger occurs in conflict zones, in countries with low economic growth and high income inequality. Moreover, extreme climate conditions determine the availability and affordability of food.

Figures

A 2021 FAO report, *The state of food security and nutrition in the world*, shows that the number of undernourished people worldwide is rising again (Figure 5). In 2020, 720 to 811 million people suffered from hunger, about 118 million more than in 2019. The target of eliminating hunger from the world by 2030 will therefore not be met.

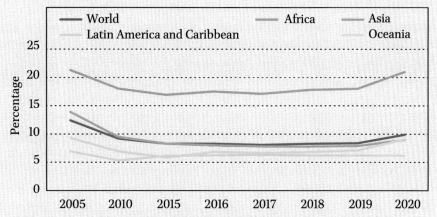

FIGURE 5   Percentage of people suffering from malnutrition worldwide. Percentages of North America and Europe are below 2.5 over the years
SOURCE: FAO, IFAD, UNICEF, WFP & WHO, THE STATE OF FOOD SECURITY AND NUTRITION IN THE WORLD 2021 (ROME: FAO, 2021), P. 11

Food diffusion—the long road from unknown to known

An old and new phenomenon

What is on our plates and in our cups at home usually feels familiar. For instance, in the Netherlands, it seems as if the following food and drink has always been there: wheat bread, potatoes, tomatoes, peppers, delicacies such as chocolate and *speculaas* biscuits, tea and coffee, with or without sugar. Our food pattern, the sum total of what we eat and drink, is visibly different from that of our neighbouring countries. This Dutch diet is like a mosaic, the individual elements—the bricks—of which originally came from everywhere, often from distant regions.

The spread of food from one area or continent to another, also known as food diffusion, is an ancient and recurrent phenomenon. About 10,000 years ago, the Middle East witnessed a gradual transition from hunting and gathering for food to cultivating plants and rearing animals as a food source. In the area now called the Netherlands, the first farmers settled in 5300 BCE on the Loess plateau of southern Limburg. They brought with them plants and animals originally cultivated in the Middle East. Farmers from this Linear Pottery Culture, the cultural period between about 5500 and 4400 BCE, cultivated a limited number of crops: the cereals 'emmer' (Figure 6), 'einkorn' and barley, as well as peas, lentils, and linseed. Poppy seeds came from the western Mediterranean. The main domestic animals kept by farmers—cow, sheep, goat and pig—originated in the Middle East.

A large-scale spread of food around the world, with major implications for the food customs of many cultures, went hand in hand with European colonisation in the sixteenth and seventeenth centuries. Spain and Portugal played a crucial role in this, bringing back from the American continent a large number of foods that are now an integral part of our daily diet. The globalisation of food thus began centuries ago. Table 2 shows the areas of origin of some important foods and demonstrates the huge influence of European colonialism on this globalisation. Notable is the huge contribution of foods cultivated by civilisations (Incas, Aztecs, Mayas) in pre-Columbian America.

From the American continent, settlers brought potatoes, sweet potatoes, cocoa, tomatoes, bell peppers, green beans, peanuts and the turkey to all sorts of places in the Old World. These foods are now an everyday part of the diet in many countries, and not just in Europe. In parts of Africa, traditional millet and yam were gradually (partly) replaced from the sixteenth century onwards by maize and cassava, relative newcomers from across the Atlantic. They are now no longer seen as

FIGURE 6   Triticum dicoccum, Emmer wheat. On a larger scale, this crop is now grown only
in Ethiopia, Turkey and Syria, where it also occurs in the wild.
SOURCE: WIKI COMMONS.

TABLE 2   European expansion in the sixteenth to eighteenth centuries and the origins of new
food and drink products that took hold in Europe

Food	Origin and plantations
green bean	tropical America
sweet potato	Central and South America
cocoa	Central and South America, Mexico
corn	Central America, Mexico
potato	South America, Andes
peanuts	Brazil
bell pepper	tropical America
chilli	tropical America
tomato	Mexico, Peru
pineapple	Brazil, tropical America
cloves, mace, nutmeg	Moluccas
cinnamon	Ceylon (Sri Lanka)
black pepper	West Java (Bantam), South Sumatra (Palembang), west coast of India (Malabar)
tea	China, plantations: Java, Brazil
coffee	Ethiopia, plantations: Java, Caribbean and South America
sugar	New Guinea, India, plantations: Caribbean, Brazil

exotic, but as an indispensable component of their own food culture. This history shows that migration, trade, colonisation and war have been the main causes of food diffusion.

Since the mid-twentieth century, we have been faced with a different dimension. In the period of modern globalisation after 1945, we increasingly see the acceptance of foods and dishes from other countries and cultures. A group of French geographers lists as the international 'taste bestsellers': hamburger, pizza, doner kebab, sushi and cappuccino. The United States has been particularly influential in popularising certain lifestyle foods. This started with the success of soft drinks, Coca-Cola, Pepsi-Cola and their derivatives, followed by the fast food culture and accompanying restaurants.

Fast food

The spread of the fast-food chain, with McDonald's as its symbol, generated resistance. When McDonald's opened a restaurant in Rome in early 1980 on the famous Piazza di Spagna of all places, protests were rife. For many Italians, it felt as if their food culture had been targetted. In France, there was action against McDonald's later, in 1999. Activist farmer and politician José Bové, together with local farmers, destroyed a branch of the US company that was about to open in the town of Millau in the department of Aveyron, in protest against the 'threat' to French food culture by multinational companies. He saw the disappearance of small farmers in France as an effect of this internationalisation.

Evidently, the path from unknown to known food did not happen automatically: there was often resistance to novelty. McDonald's itself, incidentally, knew that flexibility was the key to handling resistance. So in India, given the cow's sacred status among Hindus, the beef burger became a lamb burger. And if the need arises, McDonald's will open a *kosher* outlet as, for example, in Buenos Aires in Argentina (Figure 7). This is an example of what British historian Peter Burke calls *cultural hybridity*. In his book of the same name (2009), he points to the increasing scale of cultural interactions, and the mixing and adoption of elements from other cultures. This is certainly also taking place in the field of food. English social anthropologist Jack Goody wrote about this phenomenon in his book *Cooking, cuisine and class: a study in comparative sociology* (1982). He pointed to the menus served in international hotels, which are composed of dishes that come from everywhere and bear little relation to the food culture of the country of residence. He wondered whether this would also start to happen at the household level. Goody calls it Esperanto cuisine, comparing it to the artificial language which is made up of elements primarily from Indo-European languages but which has nothing of its own.

FIGURE 7   A *kosher* McDonald's in Buenos Aires
PHOTO BY USER GEOGAST, WIKI COMMONS

The spread of sugar and spices

Sugar and spices were expensive in the late Middle Ages. In the Netherlands and north-western Europe, they were known only among the small upper strata of society: nobility, high clergy and very wealthy citizens in the cities. The precious commodity of sugar was used as medicine. Medical textbooks of the time recommended this sweet substance as a means for the sick to regain their strength. The vast majority of people were unfamiliar with sugar and spices. Gradually, the demand for these products increased, which had huge consequences for the peoples of the Americas and Africa.

From the late Middle Ages, the modest cultivation of sugar cane shifted from the Mediterranean (including Cyprus) to the Atlantic islands of Spain and Portugal. The beautiful white sugar of Madeira and the Canary Islands was known for its beneficial properties. The early Spanish sugar plantations in the Americas imported the technical knowledge and craftsmen of the Canary Islands. Sugar cane cultivation in the Dominican Republic, in Santo Domingo, was based on slave plantations. This had dramatic consequences for the indigenous population who died en masse from imported diseases. To reinforce their labour force, plantation owners in Brazil and the Caribbean brought in people from Africa. This forced large-scale migration of Africans to the sugar plantations (and cotton plantations in the United States)

FIGURE 8   Coloured lithograph by Petit after a drawing by Théodore Bray (1818–1887). Impression
of harvesting sugar cane on a slave plantation
FROM THE COLLECTION OF THE TROPICAL MUSEUM IN AMSTERDAM,
WIKI COMMONS

is called the African Diaspora. The Netherlands played a role in this sad history.
For the sugar plantations in Suriname (Figure 8), people were taken from the Gold
Coast, now Ghana. A large number of Dutch forts on the Ghanaian coast testify
to this.

More on food and slavery

Slave labour

The incredibly rapid spread of some popular foods would have been unthinkable without slavery. Coffee production, for instance, took off after the
increasingly widespread use of forced labour. The Dutch were the first to see
the potential of the product and planted coffee on Java as early as 1696.[1] In the
eighteenth century, however, production shifted almost entirely to the New
World. By 1780, eighty per cent of the world's coffee came from the Americas,
especially from the French colony of Saint-Domingue (Haiti), which was then
exporting some eighteen million kilos of coffee a year.[2] Slave labour was the
norm on plantations in the West Indies. Over a period of several centuries,

about 12.5 million Africans were forced to cross the Atlantic; approx. 10.7 million of them made it across alive.[3]

The crossing

Many European slave ships departed from Fort Elmina, Ghana, where provisions and barrels of drinking water were taken on board. The journey across the Atlantic took about two months, but many slaves were already on board much earlier. The log of the ship 'De Eenigheid' (1761–1763) portrays life on board. The prisoners could barely stand or lie down in the cramped space between decks. These quarters were cleaned about once a week during the crossing and fumigated with juniper berries and incense. The prisoners' main meal usually consisted of groats, field beans and water. They had to be very frugal with drinking water. The water for boiling the groats was therefore mixed with 'one quart salt water'. The best food was eaten just before arrival: the men and women from West Africa were then usually given a lot of porridge and meat to gain strength in a short time, so they would look healthier and fetch a better price.

The price of sugar

Once they arrived, things soon got worse, especially on sugar cane plantations. On some British islands, life expectancy for those who worked on these plantations after arrival was only seven years. Some, like ship's purser Aaron Thomas, condemned these conditions. 'I never more will drink Sugar in my Tea, for it is nothing but Negroe's blood', he wrote at the end of the eighteenth century. But most Europeans did not know, or did not want to know, and sugar consumption skyrocketed. In 1773, for example, the British were already consuming 20 times more sugar than 110 years earlier. Because the use of slave labour for European food production was a centuries-old global practice, cannot be generalised, however. Working and living conditions could vary greatly, both over time, between regions, and even between plantations that were a stone's throw from each other.

On the plantations

The diet of slaves on the plantations gives an impression of their living conditions.[4] One or more slaves—usually women—cooked for the whole community. The diet on the plantations in the Americas consisted primarily of a source of carbohydrates: porridge made of maize, plantains or cassava, rice, or something similar. This was usually supplemented with a small amount of protein. On Surinamese plantations, this could be, for example, a piece of cured meat or fish. In a general sense, this food pattern somewhat resembled what was common in large parts of West Africa, where people had to rely on filling gruels of grains and tubers. Plantation owners often provided some other foods as well. In parts of North America, slaves sometimes received sweet potato, molasses (sugar syrup), buttermilk and various types of beans. As for fruits and vegetables, a more important supplement came from the gardens that the slaves themselves maintained. Furthermore, they often caught game, and if there was water, they could fish. Archaeological research on French Martinique shows that slaves kept pigs and chickens, caught rats or possums, and fished for crabs and seafood. They sometimes sold such products at market. This provided some extra income as well as social contact with peers from other plantations and freedmen or women.

Food supply

Various historians believe that slaves regularly had enough food, or 'fuel', available to them, but that their diet was very unvaried. Local differences, however, are significant. In the colonies in the Americas, including Suriname, slaves regularly fled because of hunger. They also died, not infrequently, of malnutrition. Workers who were less 'productive', such as the elderly and children, were especially poorly fed. Incidentally, administrators had drawn up rules to regulate this food supply. From 1685 until the abolition of slavery in 1848, the French colonies had the so-called *Code Noir*, which prescribed a weekly minimum of carbohydrates and protein. However, there was poor enforcement of these provisions. It is therefore not surprising that slaves started stealing, either from their own master or from other plantations. There is evidence that stealing from one's own master had an important symbolic meaning: it was a way to rebel against slavery and undermine the plantation owner's authority. Successfully stealing, preparing and eating a pig could be an important point of pride.

A new cuisine

Maize, pumpkin and other New World crops were combined by slaves with staples from West African cuisines, such as okra (a vegetable) and sesame seeds. As such, they created entirely new cuisines in different places. Typically, everything was simmered slowly in one pot, as can still be seen in the complex dishes called *gumbo* eaten in the United States and elsewhere (Figure 9). More specific dishes were also popularised by these people: the *Benne* waffles (after the Bantu word for 'sesame') in South Carolina, for example. Or *akara* balls, deep-fried fritters made from beans, which quickly sold well on the streets of Brazil. Small wonder that many words for food and drink in the Americas can be traced back to African languages.

In a general sense, it can be assumed that the food that slaves ate together despite, or perhaps partly because of, the often poor conditions was of great cultural significance to them—in part because the food could represent a very tangible connection to the life they had in West Africa and was therefore a way

FIGURE 9   A gumbo with chicken, ham and sausage
             PHOTO BY USER FÆ, WIKI COMMONS

of expressing their own identity. Moreover, communal meals were the main daily social occasion, where quite conceivably a sense of community could emerge in an often violent environment.

Sources

1   Jonathan Morris, *Coffee: A Global History* (London: Reaktion Books, 2019) 66.
2   Richard W. Franke, 'The Effects of Colonialism and Neocolonialism on the Gastronomic Patterns of the Third World', in: Marvin Harris and Eric B. Ross (red.), *Food and Evolution* (Philadelphia: Temple University Press, 1987) 455–79, 456; Morris, *Coffee*, 82.
3   Matthias van Rossum and Karwan Fatah-Black, 'Wat is winst? De economische impact van de Nederlandse trans-Atlantische slavenhandel', *Tijdschrift voor Sociale en Economische Geschiedenis* 9:1 (2012) 3–29, 8, 14–15.
4   The part about the food eaten by slaves is based mainly on the following sources: Lizzie Collingham, *The Hungry Empire: How Britain's Quest for Food Shaped the Modern World* (London: The Bodley Head, 2017); Herbert C. Covey and Dwight Eisnach, *What the Slaves Ate: Recollections of African American Foods and Foodways from the Slave Narratives* (Santa Barbara, CA [Etc.]: Greenwood Press, 2009); Rachel Laudan, *Cuisine and Empire: Cooking in World History* (Berkeley, CA [Etc.]: University of California Press, 2014); Sidney W. Mintz, *Tasting Food, Tasting Freedom: Excursions into Eating, Culture, and the Past* (Boston, MA: Beacon Press, 1996); Linda Newson and Susie Minchin, 'Diets, Food Supplies and the African Slave Trade in Early Seventeenth-Century Spanish America', *The Americas* 63:4 (2007) 517–50; Alex van Stipriaan, *Surinaams contrast: Roofbouw en overleven in een Caraïbische*

plantagekolonie, 1750–1863 (Leiden: Brill, 1993); Diane Wallman, 'Slave Community Food Ways on a French Colonial Plantation', in: Ken Kelly and Benoit Bérard (red.), *Bitasion: Lesser Antilles Plantation Archaeology* (Leiden: Sidestone Press Academic, 2014) 45–68.

No longer a luxury

The demand for spices was one of the reasons for establishing the United East India Company (VOC) in 1602, to try to break the Portuguese monopoly on the spice trade. This eventually led to the Dutch colonisation of much of the Indian archipelago, which only ended after a fierce war of independence after World War II.

Black pepper and the so-called fine spices—cloves, mace, nutmeg and later cinnamon—were extremely profitable products for the VOC for a long time. The spices were generally used in small quantities: as flavouring, for preservation or in cosmetics. Cinnamon, for example, is one of the main ingredients of *speculaas* biscuits, now seen as an archetypal Dutch delicacy (Figure 10). But this product also contains other 'mixed spices', such as nutmeg, white pepper, aniseed, ground ginger, cardamom and coriander. The centuries-old popularity of cake in our food culture, with the highlights being the Groninger and Deventer cake, can be traced back to the spice trade of the VOC.

By the second half of the nineteenth century, sugar and spices were now no longer unusual luxuries. Large-scale sugar cane production in the Dutch East Indies and the cultivation of sugar beet in the Netherlands made sugar ever cheaper. The production of pepper and cinnamon was expanded. But still, sugar was not cheap in the Netherlands. In 1902, the 'Anti-Sugar-Excise Union' was even founded with the aim of getting the relatively high excise duty on sugar reduced and thereby promoting

FIGURE 10   Speculaas biscuit tin from NV Delicatessenfabriek of the Maison van Asperen, probably 1890–1930
SCAN: ALF VAN BEEM, WIKI COMMONS

the use of sugar. In 1910, medical professor Cornelis Pekelharing (1848–1922), one of the founders of vitamin theory, also called for a reduction in excise duty on sugar. Pekelharing was of the opinion that sugar was not, as he called it, an 'adequate' foodstuff, but that it nevertheless provided necessary carbohydrates to workers and was therefore an important source of energy.

This sugar propaganda, by the way, runs counter to the modern advice of the Dutch Nutrition Centre (Voedinggscentrum.nl). According to this institute, sugar certainly provides energy, but no other useful nutrients. Moreover, adding sugar to foods provides more calories than necessary and increases the risk of obesity. The Nutrition Centre advises moderation with sugar-rich products, such as soft drinks, sweets, biscuits and cakes. Once a luxury product, sugar is now seen as a threat to public health.

New thirst quenchers—coffee and tea

Thirst quenchers are very important for nutrition and health. Water is the most important nutrient for humans: without a sip of this liquid, we can survive only a few days at most. In the Middle Ages, beer was the most important thirst quencher.

In the cities there were breweries, while in the countryside brewing was largely done in-house. And it was women's work, by the way. People knew from experience that water from streams, rivers and canals was unhealthy, whereas beer appeared to be relatively safe. Today, we know that water can be contaminated with pathogenic microorganisms that are rendered harmless by the boiling process during brewing.

In the Netherlands in the Middle Ages, the now popular drinks of coffee and tea were unknown. It was not until the nineteenth century that beer was gradually replaced by these drinks. There are national differences in terms of preference for coffee or tea: the Netherlands, Germany and France are coffee-drinking countries, while in England, more tea is consumed. So why have coffee and tea become so important in our food culture? One relevant aspect is the slightly stimulating effect of the caffeine in both drinks.

A major change in brewing technology may also have played a role. For a long time, beer was cloudy and had a low alcohol content. The latter property made it a good thirst quencher. However, after 1860 a different brewing method was introduced in the Netherlands. The Bavarian method, which is based on low-temperature fermentation, produces a clear, but also stronger beer. It has a good flavour, but is less suitable for drinking in large quantities due to the higher alcohol content.

Tea and coffee thus became a good alternative to beer. By the sixteenth century, tea, coffee and cocoa had reached Europe at about the same time. The new beverage of chocolate milk made with cocoa from Mexico did not function as a remedy for thirst. Yet for a long time, it was seen as a tasty drink that was uplifting for both healthy and sick people. No wonder the Droste cocoa advertisement from around 1900 shows a nurse with a fragrant cup of chocolate milk on her tray.

Coffee

Coffee originates from the Ethiopian region of Kaffa. The word coffee comes from the Arabic *qahwa* which was changed to *kahve* in Turkish. The Dutch word 'koffie' was adopted by the English and became 'coffee'. Ethiopia has extensive coffee ceremonies. Coffee from Yemen was probably finally introduced to Istanbul via Egypt in the 16th century. Incidentally, the Islamic world was initially sceptical about coffee. Islamic scholars and imams wondered whether, like alcohol, it was an addictive substance, or a mere beverage. Over time

Islamic scholars decided that coffee did not contradict the principles of the Quran. For Muslims, who often do not drink alcohol, this made coffee an attractive alternative to spirits.

The Ottoman Empire had a significant influence on the spread of coffee as a beverage. With their conquests in the Balkans, the Ottomans brought coffee with them. So, after the Turkish siege of their city in 1683, the inhabitants of Vienna discovered that coffee was a delicious beverage. From Austria, coffee spread among the aristocracy and well-to-do bourgeoisie of the Habsburg Empire. In Vienna, the typical *Viennese coffee house* emerged, and became a meeting place for businessmen, artisans and politicians, equipped with a reading table with newspapers and magazines (Figure 11). Coffee was a luxury product for a long time. It was not until approx. 1900 that this drink became more widely consumed. Middle-class breakfast

FIGURE 11   Café Griensteidl in Vienna in 1896. Opened in 1847, the Coffee House was a meeting
point for artists.
PAINTING BY REINHOLD VÖLKEL (1873–1938)—STADTCHRONIK WIEN, VERLAG
CHRISTIAN BRANDSTÄDTER P. 360, WIKI COMMONS

in central Europe then often consisted of coffee with milk and a roll with butter and jam.

In Hungary, the integration of coffee into the culture took a lot longer. It was accepted only after the Ottomans were completely driven out of Hungary in the early 18th century. During the Siege of Vienna, the area now known as Hungary had been part of the Ottoman Empire for 150 years. Because coffee drinking, in the eyes of Hungarians, was a custom of the occupying forces, they were not interested in the drink for a long time. In both Hungary and Transylvania, the act of drinking coffee took on a symbolic meaning during this period. Drinking coffee Turkish style, i.e. sitting on the floor on a cushion, was Islamic, while sitting on a chair at the table with a glass of wine was seen as Christian.

Coffee in the Netherlands

In the Netherlands, the first mention of the sale of coffee as a beverage comes from Amsterdam in 1663. From the eighteenth century on, the coffee house became a well-known phenomenon in Dutch cities. Unlike tea, coffee was for a very long time a drink that could only be consumed in public places. After all, preparing coffee was not easy, you couldn't just do it at home. The coffee beans first had to be roasted and then ground finely in a mortar. On a hot plate there was a large jug of hot water that was poured into a cup with the ground coffee. Another method was to put the ground coffee in a jug of cold water and then bring it to the boil.

Initially, the Dutch depended on coffee supplies from traders from the Middle East (the Levant) and from Greeks and Armenians. Later, the VOC itself traded coffee beans from Java. From 1723, the West India Company obtained coffee from Suriname, where slaves worked on coffee plantations. Coffee quickly became cheaper as a result. Medical professionals at this time complained, incidentally, that coffee consumption by the working class was leading to a poor work ethic.

In the nineteenth century, the sale of roasted coffee beans in shops, as well as household coffee grinders and jugs, made brewing coffee increasingly easy. As a result coffee became a home beverage. Moreover, the supply of coffee improved due to the cultivation system adopted in Java (1830–1870). This system involved forcing Javanese farmers to plant one-fifth of their land with crops for the colonial administration: coffee, sugar and plants providing the blue dye indigo (Figure 12). It yielded a lot of money for the Dutch treasury and relatively cheap coffee for the Dutch consumer. It is open to debate how heavily the cultivation system weighed on Javanese peasants. Multatuli, pseudonym of Eduard Douwes Dekker (1820–1887), clearly took sides with his book *Max Havelaar, or De koffijveilingen der Nederlandsche Handelmaatschappij* (1860). The book played a major role in the fight for the abolition of the cultivation system.

By Multatuli's time, coffee and tea had become part and parcel of the Dutch diet. By the end of the eighteenth century, coffee was already fully established in the city and the countryside. When real coffee was too expensive, people used coffee substitutes, such as the roasted and ground root of the chicory plant. Coffee also

FIGURE 12   Indonesian women sorting coffee beans, circa 1905 in Medan
             PHOTO BY CARL JOSEF KLEINGROTHE. DIGITAL COLLECTION LEIDEN
             UNIVERSITY LIBRARY

FIGURE 13   Mrs J.M. van Walsum-Quispel, wife of Mayor Mr G.E. van Walsum (next to her), opens
             the espresso bar of Engels restaurant on Coolsingel. 25 September 1957
             ROTTERDAM CITY ARCHIVE

became a remedy for alcohol abuse in the nineteenth century. Proponents of banning alcoholic beverages set up coffee houses at factories and ports. Workers who would otherwise go to the pub to drown their wages could go there for a cup of coffee and some company. Well-known were the coffee houses of the Temperance Society Against Alcohol Abuse, founded in 1875.

At the end of the nineteenth century, coffee was increasingly assigned to specific times of the day: a cup of coffee in the morning and after the evening meal became the custom. Much coffee was drunk at home, while the Italian espresso and cappuccino remained associated with cafés and restaurants. In Amsterdam, espresso was available for the first time in three espresso bars during 1956. It was a special event that drew the attention of the media. The wife of Rotterdam's mayor, Mrs J.M. van Walsum-Quispel, who opened the espresso bar at the Engels restaurant on Coolsingel in Rotterdam in September 1957, featured prominently in the photo (Figure 13). *De Volkskrant* newspaper counted over 300 espresso bars spread across the country by the end of 1957. A final step in the history of coffee as a home beverage in the Netherlands followed at the end of the twentieth century: the introduction of espresso machines that were easy to use at household level meant that espresso and cappuccino could be made in the home.

Tea

Because of the nature of the product—dried tea leaves—tea was a home beverage from the outset. It could be quickly brewed by pouring boiling water into a pot over a spoonful of tea leaves. Today, tea is the second most consumed beverage in the world after water. It originated in China. How long people have been drinking tea there remains a matter of debate. According to Chinese lore, tea has been around as a beverage for about 4,500 years. Under the Song dynasty, a thousand years ago, tea was a beverage consumed by everyone in the Chinese empire. The Portuguese, Dutch, English and Spanish learned about this beverage through their contacts with China.

The first small quantity of tea was brought to the Netherlands around 1610. Initially, it was viewed only as a curio. The VOC played an important role in the spread of tea in Europe, with the first auction in Amsterdam in 1651. Around 1680, tea began to become popular in Europe not only as a stimulant, but also as a medicine. Early on, doctors reacted very positively to this new *herb*. The thinking was that it helped purify the blood, among other things. Moreover, drinking tea would promote well-being and was a remedy for all kinds of ailments of the mouth, throat and stomach. The most famous proponent of tea consumption was the physician Cornelis Bontekoe (1647–1685), pseudonym for Cornelis Dekker, a personal physician to Elector Friedrich Wilhelm I of Brandenburg. Bontekoe published a treatise in his *Tractaat van het excellenste kruyd thee* (1678) in which he described tea as a cure for all ills. It was even said to prolong life. By the way, it was rumoured that Bontekoe was paid by the VOC for promoting tea in this way. He introduced in

FIGURE 14   Working on the Naga Hoeta tea plantation near Pematangsiantar in Sumatra, c.1905
PHOTOGRAPHER: CARL JOSEF KLEINGROTHE, LEIDEN UNIVERSITY LIBRARY
DIGITAL COLLECTION

Brandenburg not only tea, but also coffee and chocolate. These drinks were touted there as an alternative to beer and to combat alcohol abuse. However, the use of tea was also criticised. The famous physician Herman Boerhaave (1668–1738), for instance, believed that the stomach and intestines were weakened by watery, hot drinks. He saw more value in good, nutritious beers.

Due to the large supply, tea prices fell. By the second half of the eighteenth century, tea could be drunk in almost all strata of the Dutch population. It became fully established as a beverage like any other. Around the middle of the nineteenth century, the Dutch grew tea on a small scale in Java. With the introduction of other varieties and English production methods, tea culture in the former Dutch East Indies experienced a period of great prosperity a few decades later (Figure 14).

In England, tea took on a special function for the working class. With milk and sugar and a piece of white bread during working hours, this hot drink became a substitute for the hot meal for factory workers. Tea was essential in this monotonous menu as it provided some warmth and comfort. For the upper classes, tea acquired an entirely different function: it took on a central role in the ritual of *high tea*, an important social gathering.

The battle for the thirsty throat

When coffee and tea imports picked up after World War II, consumption rose rapidly. But the now traditional drinks of tea and coffee were facing competition from Coca-Cola and other soft drinks. Coca-Cola had a modest presence in the Netherlands before the war. Until the 1950s, soft drinks were a luxury item for most consumers and were only consumed at special events. At home, people drank lemonade: a little syrup mixed with tap water. Immediately after the war, Coca-Cola began advertising on a large scale and the soft drink quickly rose in popularity. There was, as it was called in the advertising world, a battle for the thirsty throat between producers of tea, coffee, soft drinks, milk, and beer. Local soft drink brands disappeared during these years in favour of US *multinationals* Coca-Cola, Pepsi-Cola, and Fanta. Exota lemonade, one of the oldest Dutch soft drink brands, disappeared. Before World War II, advertisements for this lemonade could still be seen everywhere on enamel signs at stations and along roadsides. Even beer brewers felt threatened by these imposters and started a collective advertising campaign in 1949: 'Beer is best again'. The dairy industry started a collective advertising campaign in 1958 to promote milk drinking, called the Milk Brigade. This battle for the thirsty throat continues to this day, as can be seen by the huge rise in availability of (carbonated) mineral waters with or without flavour.[1]

Interestingly, the required temperature for pouring drinks has not remained constant over the past centuries. Two major changes took place. First, from the middle of the eighteenth century, lukewarm, traditional beer was gradually replaced by the hot drinks of tea and coffee and to a lesser extent by chocolate milk. Then, in the twentieth century, from the 1960s onwards, chilled drinks, especially soft drinks, became popular. With the development of modern refrigeration technology, it was already possible by the end of the nineteenth century to serve chilled beer in pubs and restaurants, but the rise in the number of households with a fridge after World War II greatly facilitated this change. Supermarket chain Albert Heijn boosted the introduction of the fridge through its Premium of the Month Club (PMC) savings system. In exchange for saved stamps, customers could buy a fridge very cheaply. While in 1962 only 12 percent of Dutch households owned a fridge, by 1970 73 percent of Albert Heijn customers already had a fridge in their home.[2]

The potato: from botanical curiosity to staple food

The potato is the indispensable ingredient in what is now called the traditional Dutch meal: the *AVG* or 'meat and two veg'. When exactly the potato arrived in Europe is not known, but the first description dates back to 1536. Just over half a century later, in 1593, the botanist and physician Carolus Clusius, Charles de l'Ecluse (1526–1609), brought his large plant collection to Leiden. A year later, the Zeeland patrician and botanist Johan van Hoghelande (1546/58–1614), a friend of Clusius, cultivated the potato in the Hortus of Leiden University. Clusius would eventually play an important role in spreading plants, including the potato and the tomato, to other gardens in the Netherlands and north-western Europe. Through the botanical

---

1   In 2015, even Exota soft drinks ('artisanal lemonades') returned to the market: 'the memories of the past together with the taste of the present'.
2   Premium of the Month Club (PMC). Source: Albert Heijn Heritage.

gardens of landowners and herbalists, part of the agricultural population soon
became aware of the existence of the new foodstuff (Figure 15). Gardeners started
growing potatoes as vegetables in the suburbs. By the end of the eighteenth century,
the potato was already the staple food in the Netherlands and thus had long since
ceased to be a botanical curiosity.

Why did the potato gain such a prominent place in our diet and how did that
acceptance take place? A number of factors played a role in this process. First, food
security was important, i.e. the availability and access to sufficient food throughout
the year. In agricultural communities with a high degree of self-sufficiency, secur-
ing or promoting food security was (and still is) vital. In cities, an additional factor
was the affordability of food. In the eighteenth century, Europe faced a food crisis,
as grain production at the time could barely keep up with population growth. As a
result, food prices rocketed. The cultivation of the new crop, the potato, had a great
advantage over grain crops such as wheat and rye. Under the conditions of the time,
the same area of arable land produced a quantity of potatoes that could feed two
to three times more people than the grain yield. Another advantage for the small

farmer and farm workers who had a piece of land was that growing potatoes did not require an expensive plough and horses. All you needed was a spade. So it was not luxury but hard necessity that caused people to switch to growing and eating potatoes. Bread was expensive because of the state excise on flour, a tax on grinding flour. This tax had been introduced during the Eighty Years' War and was not abolished until 1855. All this made the potato an affordable alternative.

Of further significance was that the new food was easy to fit into existing preparation methods and dietary patterns. No radical changes were needed. After all, in our regions people had long been familiar with *potagies* or stews of edible tubers and root vegetables such as parsnips. Tubers could be replaced by potatoes without too many adaptations. Another factor that promoted the popularity of the potato was that it was relatively easy to use in the existing agricultural system. Since the Middle Ages, the three-field system had been in use, with a field growing winter wheat in the first year, summer wheat in the second year and left fallow during the third year. After the transition to the four-field system, grains were alternated with turnips, clover, root crops and grasses, the 'fallow crops'. Potatoes could also easily be grown as one such fallow crop.

Despite all these advantages, the potato was not immediately widely accepted as a food. Because of its resemblance to the native nightshade, a poisonous plant, the potato plant was highly suspect. All kinds of diseases, such as fevers, leprosy and tuberculosis, were attributed to eating potatoes. The prejudice gradually disappeared by the end of the eighteenth century. In the decades that followed, however, the medical profession instead pointed to the unvaried potato diet as the cause of the poor health of the lower classes.

Utrecht professor Gerit Jan Mulder (1802–1880), physician and chemist (Figure 16), wrote full of concern about the unvaried diet of the poor, who ate mainly potatoes. In his 1847 pamphlet, *De voeding in Nederland, in verband tot den volksgeest*, he warned against the lack of sufficient protein-like substances in the diet. This, he said, would undermine the labour power of the working population and would be one of the causes of the poor's lack of physical and intellectual strength. For children, a diet consisting only of potatoes would lead to mortality or, if left to live, illness and weakness after their miserable youth. Moreover, the lack of stimuli due to the unvaried potato diet would lead to alcohol abuse.

The potato nevertheless had positive effects on public nutrition. Thanks to its affordability, people could eat a basic meal on a minuscule income during the period of the industrial revolution. Moreover, the potato is very rich in vitamin C, which was very important in winter, and while its protein content compared to wheat is lower, it is of good quality.

One problem in the early stages of potato cultivation was monoculture: year in, year out, potatoes were grown in the same field without crop rotation. This made the potato vulnerable to diseases. The infamous crop failure of 1845, due to a fungus (*Phytophthora infestans*), caused famine beyond the borders of Ireland. This potato

FIGURE 16
Gerrit Jan Mulder. Portrait by Johan
Heinrich Neuman (1819–1898)
UNIVERSITY MUSEUM COLLECTION,
UTRECHT, WIKI COMMONS

blight also caused much starvation and misery in the Netherlands in the period
1845–1849. Professor Evert W. Hofstee (1909–1987), one of the Dutch pioneers in
social and historical demography, estimated the number of victims in that period
at about 53,000 out of a total population of 2.4 million. This period of poverty and
hunger caused great social unrest that led, among other things, to food riots.

The Groningen professor of physiology, Derk Huizinga (1840–1903), wrote in
1882 that the European starvation diet of potatoes and gin bred people who at the
age of 40 looked like they were 60. But as a result of the industrialisation of the
Netherlands, general prosperity did slowly increase after 1870 and bread regained
its place in the diet of the lower classes. The potato, however, did not disappear
from the diet and continued to develop into a core part of the Dutch hot meal. At
the end of the nineteenth century, what is now seen as the traditional Dutch hot
meal came into being: potatoes, meat and vegetables.

It was not until a rise in prosperity, after 1950, that a decline in potato consumption began. Incidentally, this also applied to other foods that had become relatively
cheap, such as bread and milk. According to per capita consumption statistics, the
average inhabitant of the Netherlands ate over 350 grams of potatoes a day in 1950.
At the beginning of the 21st century, that was only about 230 grams. The meat, veg
and potato diet is under pressure. Will young consumers' preference for pasta and
rice cause our potato to disappear from hot meals? Probably not, but it is certain
that the potato will increasingly have to share its place with other staple foods.

The integration of the tomato

The name tomato is derived from the Aztec word *xitomatl*. After the conquests in Mexico of the Aztecs, the tomato was brought to Europe by the Spanish. Compared to the potato, the introduction of the tomato as a food in the Netherlands took much longer. The inventory of the Hortus of Leiden University shows that the tomato, like the potato, was already present in 1594. However, this fruit, whose yellowish variety was the first to become known in Europe, was seen as an ornamental crop, or even a 'love apple'. The Italians thought the yellow tomatoes were so beautiful, they called them *pomi d'oro*, golden apples.

There is a recipe for tomato soup in an Italian cookbook dating as far back as 1692, but it took longer for tomatoes to really become part of Italian cuisine. Tomato sauce with Italian pasta became a widespread phenomenon. In nineteenth-century Hungary, the national dish became goulash: beef with peppers and tomatoes. In the early twentieth century, the food industry promoted the use of tomatoes by producing canned tomato soup and small cans of tomato paste (Figure 17).

In the Netherlands, the tomato was for a long time an ornamental crop rather than a foodstuff. It was grown in some country houses and by garden lovers for its appearance, but not for eating. Red and yellow tomatoes were left hanging as long as possible until they started to rot. As far as is known, growing tomatoes for consumption did not start until around 1895 in a grape greenhouse in Poeldijk. These tomatoes were destined for export to Germany and England and for a few specialist shops in the Netherlands.

The introduction of the tomato as a food in the Netherlands was quite difficult until the horticultural sector began marketing tomatoes. A major improvement in urban vegetable supply was the establishment of horticultural auctions. In 1913, twenty-two of these associations decided to work in close cooperation. Four years later, this led to the establishment of the Central Bureau of Horticultural Auctions

FIGURE 17   Canned tomatoes from the Happy Home Brand in the US *c.*1920
ILLUSTRATOR UNKNOWN—MUSEUM OF HISTORY AND INDUSTRY COLLECTION,
WIKI COMMONS

FIGURE 18   Stand displaying tomatoes at an exhibition in Ridderkerk, 2 January 1927
            SOURCE: REGIONAL ARCHIVE, DORDRECHT

in the Netherlands. This bureau went on to play an important role in promoting fruit and vegetables among the population. An advertising fund was set up in 1924. Inspired by experiences in the United States and England, collective advertising campaigns were initiated for fruit and vegetables. As the first product for this collective advertising, the agency chose the tomato. The campaign was wide-reaching with 7,500 coloured advertising plates bearing the slogan 'eat tomatoes, healthy for young and old' and 250,000 booklets with tomato recipes. There were also exhibitions presenting the new product (Figure 18).

In the Netherlands, the quantity of fruit and vegetables consumed continued to rise from the beginning of the 20th century until 1939. This was partly because the prices of fruit and vegetables produced in the Netherlands remained low from 1929 onwards because of export problems. Moreover, until the Second World War, fruit and vegetables were constantly advertised collectively (Figure 19). In 1931, the Central Bureau of Horticultural Auctions noted with satisfaction that from a scientific perspective, fruits and vegetables were now considered indispensable for human health because of their high vitamin content. Partly as a result, a successful collective advertising campaign to get rid of tomato surpluses, including in the form of canned tomato soup, was carried out with government support during the economic crisis of the 1930s. Annual tomato consumption doubled during this period

FIGURE 19   Fruit from Holland's soil is healthy! Central Bureau of Horticultural
Auctions in the Netherlands, 1920–1930
ROTTERDAM CITY ARCHIVES

to about two kilos per person. Subsequently, World War II did not signal a break
with the past: tomatoes became a regular part of the diet. In addition to fresh tomatoes, canned tomato soup and other tomato products, tomato ketchup now entered
the market.

Tomato ketchup is an American innovation. Sauces under the name ketchup
were already known in England in the eighteenth and nineteenth centuries, but
these were dark sauces based on mushrooms and onions. The name points to inspi-
ration from East Asia, where salty and sour sauces had existed under similar names
much earlier. It was Henry John Heinz (1844–1919), the son of German immigrants,

who concocted the famous tomato ketchup by adapting a Chinese recipe for Cat Sup, in Pittsburgh. The product entered the market in 1876. Heinz is still the market leader (at 70% of sales) for tomato ketchup in the US.[3] In the Netherlands, the product has long since ceased to be found exclusively in snack bars and fast-food restaurants: there will be few households without a bottle of ketchup in the fridge these days.

Decline in fruit and vegetable consumption

In the early 1980s, the place of fruit and vegetables seemed firmly established in the Dutch diet. In a review of the consumption development of horticultural products, Wageningen professor of market science and market research Matthew Meulenberg (1931–2006) wrote that in that period, health was increasingly experienced as a necessary condition for functioning well in society. As a result, the importance of nutritional value when assessing food increased. Other factors considered to influence consumer behaviour were the need to save time in preparing meals and environmental awareness that reinforced the preference for 'natural' products. However, the food surveys revealed a declining trend in vegetable consumption. Roughly speaking, it can be said that vegetable consumption declined among consumers under 55 years of age. Given this trend in vegetable and fruit consumption, the horticultural industry was very interested in the potential health effects of vegetables and fruits on which there is scientific consensus. They are a source of fibre, vitamins and minerals and are relatively low in salt and fats. Fruit and vegetable consumption has been associated in epidemiological studies with lower risk of various chronic diseases, such as cancer, cardiovascular disease and diabetes. However, the underlying mechanisms are less well known.

More on fruit and vegetable consumption

Adel den Hartog wrote the above in 2010. We still don't know exactly why fruit and vegetables are healthy. Research is being done, but there has been no breakthrough as yet. High vegetable consumption appears to be associated with subtle effects on energy metabolism, inflammatory processes and the degree of oxidative stress in the body.[1] Because it appears difficult to demonstrate which substances vegetables and fruit contain and what the health effects of those substances are, it is not yet easy for producers to make health claims. Another angle is to look at the possible connection between certain diseases and the composition of the bacterial population in the gut that is

---

3   Heather Haddon & Annie Gasparro, 'The New Shortage: Ketchup Can't Catch Up', *Wall Street Journal*, 5 April 2021.

influenced by our diet. It has been shown that consumption of a lot of fruits and vegetables contributes to a favourable bacterial composition, mainly thanks to the fibres in these products.[2] Greenport Venlo and Grow Campus started the project 'The Values of Fruit and Vegetables' in 2019 to further investigate these issues.[3]

Problematic figures

Convinced of the healthiness of fruit and vegetables, Dutch government agencies are concerned about declining consumption. The latest RIVM (the Dutch National Institute for Public Health and the Environment) poll shows that few people consume the recommended amounts (Figure 20). Between 2012 and 2017, only 16 per cent of adults ate 'enough' vegetables (200 grams per day), and only 13 per cent met the fruit target (also 200 grams) which is part of the Disc of Five.[4] In an international context, the figures are also low: the proportion of Dutch people eating fruit and vegetables every day is well below the European average.[5] And this is despite the fact that most Dutch people rank fruit and vegetables among the healthiest foods.[6] Apparently, many of them don't believe they need to eat more of them: indeed, Dutch people give their own food choices a whopping 7.1 on average.[7]

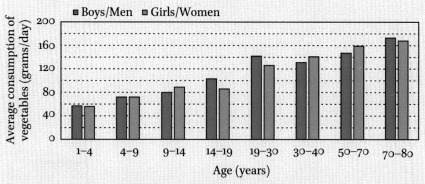

FIGURE 20   Average consumption of vegetables in grams/day by age
SOURCE: RIVM, FOOD CONSUMPTION SURVEY, REPORTING PERIOD 2012–2016

Always too little

The creator of the Disc of Five, the Dutch Nutrition Information Agency, was cautious about its advice on fruit and vegetables just after World War II. Three or four tablespoons of vegetables were enough, and the bureau did not consider fruit important at all at that time. In subsequent decades, plant-based foods became increasingly prominent in the millions of leaflets printed by this forerunner of the Nutrition Centre. Yet until the year 1975, the agency observed, vegetable consumption in the Netherlands remained 'almost constant' and thus too low.[8] Sometime later, in 1988, the first food consumption survey showed how serious the situation was: the average Dutch person ate only 144 grams of vegetables and 125 grams of fruit per day. Even vegetarians rarely followed the guidelines of the Disc of Five. Subsequent surveys, in the years 1992 and 1998, showed that consumption of both food groups had declined still further.[9]

Fruit from distant lands

Meanwhile, in the post-war period, the supply of fruit and vegetables did increase considerably. Whereas in the 1930s the tomato was still touted as a surprising addition to the menu, it was only after World War II that many new types of fruit and vegetables really appeared. Raw vegetables, which are part of the culinary traditions in Mediterranean countries, were called rabbit food by the Dutch. It was only after 1950 that eating raw lettuce and salads became more common here. In 1956, it was estimated that around 25 types of vegetables were available in the Netherlands, a number that rose to 57 in 1990.[10] A huge increase in intercontinental flights caused all kinds of exotic novelties to appear on the Dutch market. Initially, many of these were quite pricey. Avocados, when they were introduced in the 1950s, cost the equivalent of almost six euros apiece. As food transportation became cheaper and cheaper, these products became more affordable, and thus more common. In promoting new types of fruit and vegetables, the Nutrition Information Agency was assisted by commercial parties, such as women's magazine *Margriet*, which in 1970 advertised all kinds of 'fruits from distant lands', from broccoli to pineapple, and from fennel to passion fruit.[11] The housewife was also instructed on how to use the products. For instance, the 'mangga' (mango), whose flavour is a bit 'reminiscent of carrots', tastes best in fruit salads.

A policy issue

Inadequate fruit and vegetable consumption tops the list of dietary factors associated with loss of healthy life years in many Western countries, according to the *Global Burden of Diseases* project. Although fruit and vegetable consumption is increasing in many countries, it is still far below WHO recommendations. This organisation advocates a daily intake of 400 grams of fruit and vegetables. Only 27 percent of the European population eat two pieces of fruit a day, and 23 percent eat vegetables twice a day.[12]

One of the projects to increase fruit consumption is EU School Fruit. With EU School Fruit, schools receive 20 weeks' worth of free fruit and vegetables for all pupils, to encourage children to eat fruit and vegetables together. The idea is that children will taste different fruits and vegetables and thus develop their tastes (Figure 21). In the school year 2021–2022, the Netherlands participated for the 13th time in this European Union-funded scheme, for which a budget of about €5.5 million was made available. The FAO also helped in the promotion, having declared 2021 as the International Year of Fruit and Vegetables. In

FIGURE 21   A poster of the EU School Fruit campaign
REPRODUCED WITH THE PERMISSION OF THE EU SCHOOL FRUIT AND
VEGETABLE PROGRAMME.

this way, the organisation hoped to raise social awareness and draw political attention to the health effects of fruit and vegetables. It also wanted to promote a diversified and balanced diet through fruits and vegetables.

The Dutch may have started eating many more types of fruit and vegetables but, as mentioned, overall consumption is still declining. What is striking here is the strong correlation between fruit and vegetable consumption and the socioeconomic status of consumers. This relationship was already known in 1961: for people with a high income back then, it was a lot easier to buy and eat healthy products.[13] Recently, critical researchers have increasingly emphasised that this is a task for the Dutch government. They point to the fact that 'healthy' food is becoming more expensive compared to 'unhealthy' products, partly because the government increased VAT on fruit and vegetables in 2019.[14] As part of the Prevention & Care initiative, the Dutch government wants to promote a healthy diet. However, it is debatable whether the measures mentioned—promoting the production and supply of healthy food and promoting healthy choice by consumers—will be sufficient to achieve the goal. In the 2018 National Prevention Agreement, healthy eating was only part of the obesity theme. There has been criticism from various quarters about the lacklustre agreements in the Prevention Agreement and how much they have been influenced by the food industry.

Sources

1   Nieuws voor diëtisten, 'TNO toont gezondheidseffecten van groenten aan'. The study in question is: Wilrike Pasman *et al.*, 'Nutrigenomics approach elucidates health-promoting effects of high vegetable intake in lean and obese men', *Genes and Nutrition* 8 (2013) 507–21.

2   Ernst Woltering, 'Het geheim van gezonde groente', *GFActueel* (2019).

3   Mark IJsendoorn, 'Brightlands wil gezondheidseffecten groenten en fruit meten', *Expertisecentrum Voedingsmiddelenindustrie* (2018).

4   RIVM, 'Wat eet Nederland: groente/fruit'.

5   Eurostat, 'Frequency of fruit and vegetables consumption by sex, age and educational attainment level'.

6   Eighty percent of people put vegetables in their top three healthiest products; 69 percent put fruit in the top three. Network for Food Experts (NVVL) and GfK, 'Voedseltrends' peiling (2015).

7   NVVL and Gfk, 'Voedseltrends' peiling (2015).

8   Voorlichtingsbureau voor de Voeding, 'De Maaltijdschijf: Kaderbrochure voor hen die voedings-voorlichting geven' (*c.*1987). This article concerns the period 1950–1975.

9   K.F.A.M. Hulshof, C. Kistemaker and M. Bouman, 'De consumptie van voedingsmiddelen naar locatie over een periode van tien jaar. Resultaten van drie voedselconsumptiepeilingen: 1987–1988, 1992 and 1997–1998', TNO (1998).

10   This is what Adel den Hartog himself concludes on the basis of several sources. Adel den Hartog, 'The changing place of vegetables in Dutch food culture: the role of marketing and nutritional sciences 1850–1990', *Food & History* 2:2 (2004) 87–103.

11   'Vruchten uit verre landen: 'n voorproefje op uw vakantie!', *Margriet* (1970) nr. 12, 104–106, 108.

12   Nicolas Guggenbühl, 'Europese consumptie van groenten: België koploper', *Food in action* [2019].

13   Dora van Schaik, 'Maatschappelijke invloeden op het voedingspatroon', *Voeding* 22:3 (1961) 95–107.

14   Jaap Seidell and Jutka Halberstadt, 'Gezonde voeding heeft een prijs', *Het Parool*, 23 June 2019; Centraal Bureau voor de Statistiek, 'Prijs voeding met 18 procent gestegen in tien jaar', 28 May 2021.

Food taboos, food laws and rules

Food taboos

Sometimes people have an arrangement with each other not to eat certain foods, even though on the face of it they have nothing against these foods. Such food bans are often called food taboos. Taboo food is food that is not consumed for religious or cultural reasons. The word taboo comes from Polynesian languages and means sacred or forbidden; it has a magical-religious meaning. Explorer James Cook (1728–1779) introduced the word into English in the late 18th century. It then became known in anthropological literature in the second half of the nineteenth century.

A food ban is subject to sanctions. The ban is enforced sometimes in small, sometimes in large groups, within a family, community or even within a society as a whole. Anyone who does eat the food is punished. This may be just a reprimand, where the person in question is ridiculed or punished. But it can also lead to expulsion from the community to which a person belongs.

Food bans cover a large and very diverse range of foods. The reasons for these bans are not always clear. Moreover, the logic of a ban from the distant past is often untraceable today. To gain some insight into the wide variety of food bans, it is useful to distinguish between two different categories: temporary and permanent food bans. A temporary ban means that some foods cannot be eaten during a certain period of life for various reasons. In West Africa, for example, some cultures prohibit women from eating vegetables or fruit during pregnancy. A permanent ban refers to a food item that should never be eaten, such as pork for Jews and Muslims.

As British anthropologist Mary Douglas (1921–2007) writes in her classic study *Purity and danger: an analysis of concepts of pollution and taboo* (1966), published in the Dutch language ten years later as *Reinheid en gevaar*, the rules are meant to bring order from chaos, to provide clarity in the world we are part of. Following food rules means obeying God's commandments and leads to reflection.

The changing power of alcohol prohibition

Beer, wine and mead (an alcoholic honey-based drink) are the classic fermented alcoholic beverages that people consumed way back in ancient times, with mead probably the oldest among them. Around the year 800, a real revolution took place in the Arab world: people learned to distil wine into brandy, *aqua vitae*. Distillation is a technique of separating two or more substances in a solution by evaporation. The principle is based on the difference in boiling points of these substances. The word alcohol derives from the Arabic *al-kuhul*, which denoted a fine powder produced by evaporating a solid and precipitating it again. From the 16th century, the name was used for essences obtained by distilling liquids. The word alcohol was included in all Western languages. The art of distilling probably spread across Europe from the 13th century onwards. The alcoholic drink, first used as medicine, developed into a stimulant. A relic of the original medicinal use of alcohol seems to be the fact that in all kinds of European languages we still toast to good health: 'Op uw gezondheid', 'Santé', 'Salud', 'Auf ihr Wohl', 'To your health'.

Alcohol and Islam

Religious bans on alcohol are rare examples of plant-based food bans. Enforcement of the ban varies considerably. In the Arab world, for example, the alcohol ban in force in Islam is generally controlled much more strictly than in countries in other regions such as Turkey or Indonesia. In strict Islamic countries, the police ensure that the alcohol ban is not violated.

Interestingly, before the Islamisation of Mesopotamia (most of modern-day Iraq), Syria and Egypt, beer and wine were accepted beverages. Viniculture probably originated some 8,000 years ago in the mountainous land between the Black Sea and the Caspian Sea. In Iran (Persia), a wine-growing tradition existed for centuries. Excavations in the Zagros Mountains, which border Iraq, revealed signs of wine production around 5000 BCE. Even after the arrival of Islam, the city of Shiraz was the 'wine capital' of Persia for centuries, famous for its high-quality wines that were traded to Europe (Figure 22).

FIGURE 22   Drinking wine in a spring garden. Persian miniature, c.1430
COLLECTION METROPOLITAN MUSEUM OF ART, ISLAMIC ART DEPARTMENT,
WIKI COMMONS

Beer was already a daily drink in Mesopotamia 6,000 years ago, while viniculture was introduced there only later. This low-alcohol beer was brewed from barley and Emmer wheat and was widely consumed. Wine, on the other hand, was more for feasts and religious occasions.

Islam originated among Arab peoples with nomadic traditions. The ban on alcohol, as well as the ban on pork, can possibly be explained by a certain feeling of superiority among these itinerant peoples over location-bound peasant life. Both brewing beer and growing grapes for wine were part of this sedentary existence. The fermentation of grains and grapes is a lengthy process, which does not suit the nomadic life. As a result, Nomads did not want to associate themselves with alcoholic beverages.

The prohibition of alcohol in Islam

The prohibition of alcohol in Islam stems from the following verse of the Quran:

> O you who believe! Verily, wine and gambling and idolatrous images and arrows for raffling are impurities belonging to the work of Satan. So avoid these (things). Hopefully you will prosper! (Quran 5:90)

However, in the part of the Quran that deals with Paradise, wine is highly valued. Indeed, *Sura* 74-12 speaks of: '[...] and rivers of wine, pleasing to the drinkers [...]'. And in *Sura* 83-25: 'They are given to drink of pure wine, well sealed'. The *Hadith*, the great compilation of recorded Islamic lore recorded between 800 and 850, is much clearer on alcohol. Whoever drinks wine is an infidel, and drinking wine is punishable by forty or eighty lashes.

Source
The Quran, in translation by J.H. Kramers, 1974, Islam religion.

Alcohol and Christianity

In countries like Syria and Lebanon, wine is still produced on a modest scale by Christian minorities. Wine has a very important religious significance in Christianity. According to the Bible, Jesus called his disciples together on the last night before his crucifixion. In the 2004 Bible translation, *Luke* 22:17 describes Jesus taking a piece of bread and saying, 'This is my body that is given for you.' After the meal, he took up the cup of wine and said: 'This cup that is poured out for you is the new covenant made through my blood.' This blessing of bread and wine, by the way, was entirely in keeping with pre-existing Jewish traditions. The drinking of wine in Europe is a legacy of both the Roman Empire and the Roman Catholic Church. After the fall of the Roman Empire and the wholesale loss of viniculture, monks in monasteries took up this practice again.

Viniculture around the Mediterranean

The history of viniculture around the Mediterranean clearly shows how the changing Christian and Islamic power relations determined the production and consumption of wine. In the Roman-Christian period, wine was drunk and grown in the countries of this region wherever it was agriculturally possible.

This changed radically with the spread of Islam. When the Arabs spread their religious faith in North Africa between 650 and 750, the vineyards were uprooted (and the pig herds also disappeared). The North African area, now known as Tunisia, had a wine-growing tradition dating back to the foundation of Carthage by the Phoenicians, around 3000 BCE. This ancient wine culture disappeared completely with the advent of Islam. Much later, countries like Tunisia (in 1881) and Morocco (in 1912) came under a French protectorate. The French *colons* (settlers) then reintroduced viniculture. After the independence of both countries in 1956, viniculture suffered a setback with the departure of the French. Nevertheless, wine is still produced in both Tunisia and Morocco.

The Muslim Ottomans were less strict when it came to alcohol. The viniculture and also pig farming of the subjugated Greek and Slavic peoples were largely tolerated. Turks themselves did secretly drink wine and the spirit known as *raki* in Ottoman Europe. Interestingly, Albania, the only European country with a predominantly Muslim population for a very long time, also *drink* wine and *raki*. Moreover, this country, which was part of the Ottoman Empire until 1918, still produces its own cognac, *Cognac Skënderbeu*, loved for its flavour and bouquet.

Migrants

So what is the place of wine and beer in the diet of Muslim migrants in Europe? Alcohol prohibition seems to be showing cracks. Quite a lot of Muslims in far-flung regions of the world are in practice quite lenient about the prohibition of wine and other alcoholic beverages. For liberal Muslims, the psychological threshold for drinking a glass of wine or beer is not that high. Research by Mohammed H. Benkheira on consumption patterns in France found that in the early 21st century, 32 per cent of Muslims surveyed drank at some point. Yet alcohol remains a thorny issue in this community. In some suburbs of Paris, in neighbourhoods with predominantly Muslim residents, wine, beer and spirits are not found on supermarket shelves in order to avoid conflicts with some more conservative believers.

Fasting

Fasting—abstaining from food for a certain period of time—came into use in Christian tradition in the second century of our era. The commandment to fast as a means of reflection and penance probably dates back to Pope Calixtus I (217–222). These practices were introduced in our regions, with the spread of Christianity.

Erasmus and the ban on eating meat

In a tract to the Bishop of Basel (1522), Dutch priest Desiderius Erasmus (1466–1536) opposed the ban on eating meat for 150 days. Erasmus (Figure 23) recognised that fasting combined with good nutrition could have a positive effect on spiritual life. But he also had objections. According to him, the ban on eating meat led the rich to use culinary feats to prepare very tasty fish, while poor people were greatly deprived. Thus, he turned against the bishops who were constantly introducing new 'feast days' on which work was not allowed, causing hunger among the poor. He further argued that people who were poor or weak of constitution, like himself, had a hard time if they had to go without food for so long. His tract was a harsh attack on some church regulations of his time.

FIGURE 23   Desiderius Erasmus
PRINT FROM 1601 BY FRANS HUYS, AFTER
DESIGN BY HANS HOLBEIN, RIJKSMUSEUM
COLLECTION, AMSTERDAM

FIGURE 24   Detail from Pieter Bruegel the Elder's The Fight Between Carnival and Lent (1559),
showing Lent as a severely emaciated woman
KUNSTHISTORISCHES MUSEUM COLLECTION, VIENNA, WIKI COMMONS

Preparing for Easter meant fasting for 14 days. The church also had the so-called
Ember days, quarterly periods of fasting, as well as the prohibition of meat on
Wednesdays and Fridays, and fasting on celebration days of various saints. The
comprehensive church regulations eventually included a ban on meat-eating for
140 to 150 days of the church year. Fasting meant avoiding dairy and eggs in addition
to a ban on meat consumption. This put pressure on the protein consumption of
both peasants and artisans in urban areas in the Middle Ages. In contrast, the nobility, clergy and wealthy citizens could afford plenty of fish.

With the advent of the Reformation, fasting rules disappeared in Protestant parts
of Europe. A well-known painting by Pieter Bruegel the Elder (1525/1530–1569), who
had ties to the Renaissance humanism of his time, is entitled *The Fight between
Carnival and Lent* (Figure 24). These days, the painting is interpreted as a plea
for temperance and the search for the right balance in life. The Catholic Church
has also toned down these rules over time. For example, Friday is no longer a
meat-free day. Fasting for Easter has now become more of a spiritual preparation
where some people renounce certain bad habits, such as smoking, eating and

drinking too much or watching TV. In Germany, the Lutheran Church has seen renewed interest in fasting as a means of reflection. In this case, fasting prior to Easter means consuming less.

In Buddhism, Hinduism and Islam, fasting has great significance in the religious experience. The increased number of believers belonging to these world religions in the Netherlands and other Western European countries therefore means an increase in fasting as a religious phenomenon. The fasting period within Islam, *Ramadan*, takes place in the ninth month of the Islamic year. The Islamic calendar is moon-based, so the fasting period traverses all seasons, as it were. During *Ramadan*, nothing may be eaten or drunk from sunrise to sunset. The Jewish faith also has six days of fasting. The most important of these is *Yom Kippur*, the Day of Atonement. This is Judgement Day, when believers ask God for forgiveness of sins and a good New Year.

Fasting is a means and not an end in itself. Believers use it to cut themselves off from the outside world and get closer to God, and to show solidarity with the poor. A new form of fasting has emerged from non-religious quarters, to be used as a training of willpower and reflection, an exercise in 'consumerism', or as a means of cleansing the body. This form of fasting usually takes place on an individual basis.

Culturally defined rules about nutrition

A temporary ban on food is linked in many cultures to either phases of the human life cycle, or illness. Key life stages include pregnancy, infancy and childhood, transition to adulthood and old age. These are the critical phases of life that can be surrounded by all kinds of actions to keep the person concerned healthy. Once this phase is over, the forbidden foods can be eaten again.

Pregnancy is a phase in which cultural practices play an important role. In some parts of Africa, eating earth is important. As a result, the unborn child is said to recognise his or her country. Others say it facilitates childbirth because earth makes the unborn child's skin smooth and oily. Each culture has its own rules. In a number of cultures in sub-Saharan Africa, women do not eat eggs or meat during pregnancy, as this would damage the health of the unborn child. Fish is sometimes refused as this would make the child look like a fish. Drinking a lot is often advised to advance labour, while sticky rice and plantains are discouraged for the same reason. Vegetables are often not advised because of the risk of diarrhoea. Fibrous and uncooked vegetables do indeed stimulate bowel movement.

Menstruation is a period of extra significance. In various cultures, menstruating women are considered unclean. They are not allowed to touch water or cook during the menstrual period. There are many myths about menstruation that restrict the lives of girls and women. In Nepal, for instance, they are not allowed to eat dairy or meat when they are menstruating. And in Malawi and Uganda, people believe that if a menstruating woman walks through a garden, all the plants will die. Plan International, an *NGO* formerly known as Foster Parents Plan, is committed to opening up the menstruation taboo for discussion.

During my fieldwork in the Central African Republic, as well as in other countries, I saw that children were often not allowed to eat eggs because it is believed that they will get sick from them. This may seem a strange custom at first glance, but there is a logic behind it. One of the village elders gave the following explanation: in the distant past, young children would scavenge around the houses, and if they saw a chicken egg lying around, they would eat it. This caused a shortage of eggs and thus ultimately caused a shortage of chicks and chickens. Therefore, the elders told the young children that eggs were poisonous and would give them a stomach ache. I have written about this and other customs with Wija van Staveren and Inge Brouwer in the book *Food habits and consumption in developing countries. Manual for field studies* (2006).

Permanent food bans

In general, Christianity has no permanent food bans, but there are some groups that make an exception. Seventh-day Adventists are one such special group. Their diet is predominantly lacto-vegetarian, while they also often abstain from tobacco products, alcohol and stimulant drinks such as coffee and tea. Adventists' health principles have a Biblical basis, with the Old Testament provisions around food, the Mosaic food laws (*Leviticus* 11) playing an important role. In the general Christian tradition, these dietary laws have been lifted, based on the apostle Peter's vision described in *Acts* 10:11–15 (see box).

Acts 10:11-15

11   And he saw the heavens opened and an object coming towards him, which looked like a large linen sheet tied at the four corners and lowered to the earth,

12   in which were all the four-footed animals of the earth, the wild and the crawling animals and the birds of the air.

13   And a voice came to him, Arise, Peter, slaughter and eat!

14   But Peter said, Decidedly not, Lord, for I have never eaten anything unholy or unclean.

15   And again, for the second time, a voice came to him, What God has cleansed, you must not take for unholy!

Source
Revised State Translation.

The role of food as a sign of one's identity is strongly reflected in food bans. Muslims and Jews in Europe can emphasise their identity and distinguish themselves from other groups by not eating pork. More generally, the significance of meat and fish is reflected in food bans, most of which are about animals. For instance, both the Old Testament and the Quran mention a large number of animals that are not to be eaten, whereas they do not ban fruit or vegetables. In the Quran, the pig is represented in two ways: as unclean food and therefore forbidden; and as a symbolic sign of abomination.

In Jewish religious life, food plays an important role. For Jews, the food laws, sanctified since the time of Moses, are a positive means of self-liberation in the service of God. For them, restrictions on food intake repeatedly prompt reflection on the unity, purity and perfection of God. The Old Testament books of the Bible, *Leviticus* and *Deuteronomy*, set out the Jewish dietary laws. *Leviticus* (*Lev.* 11) contains a long list of which foods are an 'abomination' (see box).

Another important aspect of the food laws is found in *Deuteronomy*: 'Thou shalt not boil the goat in its mother's milk' (*Deut.* 14: 3–21). As a result, in Jewish cooking, dairy and meat are strictly separated. Food in traditional Jewish cooking must be subject to the entirety of the food laws, the *Kashrut* (Hebrew: suitable). Food that is permitted within these laws is called *kosher* in the Netherlands and other Western countries.

> **Permitted and forbidden species according to Jewish food laws (Kashrut)**
>
> Of the mammals, ruminants with cloven hooves are allowed to be eaten. Ten animals are mentioned by name: cow, sheep, goat, deer, gazelle, antelope, ibex, badger, wild ox and the wild sheep. Animals that do not ruminate, such as the pig, are prohibited even though the animal has cloven hooves. There are also animals without split hooves that do ruminate, such as the camel and hare, which are still banned.
>
> A number of birds are listed that may not be eaten, including eagle, bearded vulture and all species of vultures, raven, ostrich, barn owl, gull, sparrow hawk, little owl, cormorant, Eurasian eagle-owl, pelican, heron and Eurasian hoopoe. Birds with a wattle, such as the chicken, are allowed. Bat is prohibited.
>
> Fish with scales and fins are allowed, but smooth fish such as eels are unclean. *Leviticus* allows the eating of four kinds of locusts. Which species of locusts are concerned is still a matter of debate. However, any living creature that swarms in water or on land may not be eaten.

Medieval Europe had no permanent food bans and thus distinguished itself in that area from Judaism with its Old Testament dietary laws. Apart from meat from different kinds of animals, fish was eaten, and snails and frogs were also on the menu (Figure 25).

FIGURE 25   Sale of fish, frogs and snails
            FROM: ULRICH RICHENTALS KONSTANZER RICHENTAL
            CHRONIK, FOL. 25R, C.1464, WIKI COMMONS

There was one exception: horse meat. This meat was taboo according to Catholic doctrine until the 19th century, because the horse was associated with pagan customs. In general, the Catholic Church was ambivalent about eating meat because it was believed to arouse 'carnal lust'. Monks were expected to feed exclusively on bread and grains. Incidentally, excavations of early Christian monasteries in Ireland show that meat was eaten anyway. Usually, however, monks tried to abide by the *Regula Benedicti*, the precepts of Benedict (*c.*480–550), the founder of the Benedictine monastic order. According to these rules, the sick and convalescent were allowed to eat meat from quadrupeds to regain their strength faster, but when they recovered they had to deny themselves of it again. To provide for their meals, the monks had a vegetable garden. Here they also grew medicinal herbs that they used for medical treatments.

CHAPTER 4

The meaning of meat

When people are served a meal that contains unfamiliar plant foods, it usually doesn't cause too many problems. Unknown fruit and vegetables are much easier to stomach than an undefined piece of meat. The emotional relationship people have with food is particularly evident with animal products. This is partly because it confronts people with the realisation that an animal has been killed so that they themselves can live.

Pork: beloved or banned

There is perhaps no meat that evokes as many conflicting emotions as pork. On the one hand, it is much loved in many parts of the world; on the other, it is seen as unclean, and is despised and reviled in many countries. The domestic pig, *Sus scrofa domestica*, is a subspecies of the wild boar, *Sus scrofa*. It is one of the first animal species kept as farm animals by farmers. Remains have been found in Jericho, one of the world's oldest cities, indicating the keeping of domestic pigs from 9,000–10,000 years ago. Excavations in China have found similar remains, some 7,000 years old.

People's ambivalent attitude towards pigs is also ancient. In Mesopotamia, present-day Iraq, no clear evidence has been found of a ban on the eating of pork. But although pigs were eaten, people did see the animals as dirty. A clay tablet found in the palace of the *Mari* kingdom (2000–1750 BCE) mentions that the pig was not clean and soiled everything. According to this text, the animal made the streets dirty and defecated in houses. At the time of the pharaohs in Egypt, pork was eaten, but there were also periods of prohibition of pork.

Beloved
In classical antiquity, pork was highly sought after by the better-off Greeks and Romans. It was so popular that collection of Roman cookery recipes, Apicius, mentions 'meat' where pork is meant. Even Pliny the Elder (23/24–79), the Roman encyclopaedist, had a great appreciation for pork. 'No animal provides more ingredients for feasting: around fifty delicacies, while other animals provide only one,' he wrote in his *Naturalis historia*. An important geographical factor played a role in the popularity of the pig among the Greeks and Romans. They had to make do with this animal partly because the Mediterranean climate of southern Europe was not suitable for large-scale cultivation of cattle. Poor people hardly ate meat anyway. Their main meal consisted largely of bread (wheat or barley) or a porridge (*puls*) of grains.

After the Arab-Islamic conquest of Egypt in 641, Coptic Christians gradually integrated the Islamic ban on pork into their traditions. That said, pork is still eaten by some Coptic Christians. In a general sense, pork is not completely absent from Christian dinner plates in the Middle East.

In Christian art, the pig is inextricably linked to Saint Anthony the Great (251–356). St Anthony, originally from Egypt, is depicted in Christian art with a pig at his side, writes art historian Silvia Malaguzzi in *Food and Drink* (2007). According to her, the meaning of this combination would be that the saint had overcome evil. Indeed, in the 15th and 16th centuries, Flemish and Dutch paintings often depicted the pig as a symbol of sin, gluttony and lust. The order of hospital brothers named after the saint, the Antonites, was founded in 1095 to combat the disease ergotism, also known as Saint Anthony's fire, and to care for plague and leprosy sufferers. The brothers were allowed to let their pigs roam freely as an in-kind payment for their nursing care. The pigs fed on the rubbish in the streets and were identifiable by a bell around their necks and marked with the T-cross. On Anthony's feast day, the animals were slaughtered, and their meat was distributed to the poor. The veneration of Saint Anthony was widespread in North Brabant and parts of Flanders.

Meat was eminently symbolic of the powerful in society. In medieval France, pork was highly prized among nobles; they consumed far more of it than peasants, who ate more beef. The nobles found beef too coarse for their delicate stomachs. Something else that determined the taste would have been the fact that beef mostly came from discarded cows, while pork came from animals reared for slaughter. Moreover, attention was paid to the pigs' feed: in autumn, they were herded in forests, where they could feast on acorns. The appreciation of pork is also reflected in a painting from the early seventeenth century that shows not only the prominent pig's head, but also pig's trotters and several strands of sausage (Figure 26).

The Melanesian inhabitants of the island of New Guinea have a strong tradition of keeping pigs. And these beasts are certainly very well cared for. Until recently, for instance, it was not unusual for women to personally breast-feed piglets that no longer had a sow! The big feasts where pigs are slaughtered en masse, stewed in a pit and eaten, are famous.

The pig is an ideal farm animal and the most widely eaten meat in the world. It is indispensable in the diet of China and Europe. China boasts the most pork lovers. In this densely populated country, pigs were the traditional source of animal protein, alongside chickens and dogs. According to the article 'Kings of the Carnivores' in *The Economist* on meat consumption in Europe, Austria has the highest consumption of pork, followed by Serbia and Spain. Denmark, ranked sixth, has shown that pigs can also have great significance in the development of modern agriculture. French researcher Anne-Elène Delavigne even wrote an article entitled 'Pas de cochon, pas de Danois' ('no pigs, no Danes'). It is a telling example of the important function of food in conveying the identity of a people or a culture.

FIGURE 26   Everything from the pig. Still life with a pig's head and other types of pork, *c.*1600–1650
by monogrammist 'JVR F.'
RIJKSMUSEUM COLLECTION, AMSTERDAM

The pig is still positively associated with prosperity. Thus, savings pots are often shaped like pigs, and the fatter the pig, the better. During the economic crisis of the 1930s, pigs were able to alleviate poverty somewhat. Agricultural workers in the Groningen countryside then set up so-called *bacon clubs*. A number of families then jointly kept a few pigs to have them slaughtered in winter. This way, people had some meat at their disposal despite low wages.

Raising pigs has great advantages over raising cattle. They are omnivorous: they eat everything and can live off waste. In that respect, they are free-range scavengers, just like chickens and dogs. On top of that, they need little space to live, which is convenient in densely populated areas. The feed conversion rate, or the conversion of a given amount of grain into a kilogram of meat, is much more favourable for pork than for beef. Overall, for beef, eight kilos of grain yields one kilo of meat, while for pork only 2.5 kilos of grain is needed.

The advent of the bio-industry with its mega-stables has significantly changed the status of pigs. To the rural population, the pig was once an economical animal that could largely be fed on leftover food scraps. With today's economies of scale, the pig is now competing with farmland and plant food intended for human consumption.

Unclean and forbidden

There are several explanations for why people consider pork unclean. Hygiene is put forward as a possible reason, because pigs scavenge in waste and root around in mud. Yet this is an unlikely explanation, as the modern understanding of hygiene dates back to the mid-nineteenth century. The possible occurrence of *trichinosis* has also been put forward as a reason for avoiding pork. *Trichinosis*, an infection of humans by the roundworm *Trichinella spiralis*, is a nasty and painful disease that can be caused by eating inadequately cooked pork. A worm that encapsulates itself in muscle tissue can be present in the body for years. In some cases, the roundworm can even reach the central nervous system. However, this explanation for the unclean image of the pig is not convincing because the relationship between *trichinosis* and eating pork was not discovered until around the mid-nineteenth century.

Geographer Frederick Simoons (1922–2022) offers a plausible explanation for the widely supported ban on pork. The core of his explanation lies in the centuries-old opposition between pastoral populations and agricultural populations that remain in a fixed place. Pigs belong to the farming population because they are unsuitable for travelling long distances in herds every day. Herders, people of the free field, however, do not have a high opinion of the housebound farmers who root around in the earth with their hands for daily subsistence. Eating meat associated with this inferior way of life would therefore be unthinkable for them. On top of that, for Muslims, pork was (and is) characteristic of the diet of Christians. Therefore, with the spread of Islam in the seventh century from the Arabian Peninsula in northern Mesopotamia, and in Syria, Egypt, and North Africa, the keeping of pigs was banned.

This history was repeated in Cairo in 2009. There, Christian minorities have traditionally engaged in manual rubbish disposal, a dangerous and dirty job. Organic waste is disposed of by Christian pig herders who make the pigs eat the city's organic waste. This phenomenon was always a thorn in the side of some Muslims. When swine flu hit in 2009, the city council decided to have all pigs culled. This was welcomed by stricter Muslims who thus saw this 'unclean' blemish disappear. For Christian pig herders, however, this was a major blow to their already marginal livelihood. Incidentally, swine flu (H1N1) is endemic among pigs, but very rarely transmitted to humans.

Halal food in the Netherlands

A study published in 2018 by the Netherlands Institute for Social Research on the religious experience of Muslims in the Netherlands focuses on nutrition. The study mainly concerns the two largest Muslim groups in our country, Turkish-Dutch and Moroccan-Dutch Muslims. A very large proportion of Muslims (almost) always eat *halal*. Among Moroccan-Dutch Muslims, 93 percent always eat *halal*, and 80 percent among Turkish-Dutch Muslims.

The proportion eating *halal* daily has grown slightly among the former group since 2006 and remained the same among the latter. Eating *halal* and being a Muslim are therefore closely linked for these groups. Among Turkish-Dutch Muslims in particular, there are also differences between generations, between men and women, and between those with a low and high level of education.

Source

Willem Huijnk, *De religieuze beleving van moslims in Nederland. Diversiteit en verandering in beeld* (The Hague: the Netherlands Institute for Social Research, 2018).

Changes

The ban on eating pork is still strong among Muslim migrants in the Netherlands and other European countries. During a geographical study tour in October 2008, organised by Utrecht University's Faculty of Geosciences, I noticed that the situation was different in predominantly Muslim Albania. From the available data and observations, I got the impression that the ban on pork was largely circumvented. Sausages made from pork were available in shops and supermarkets. The presence of large billboards advertising salami was striking, sometimes right next to a mosque. Apparently, few people cared.

Furthermore, although the ban on pork is also firmly rooted in the Mosaic dietary laws, there are signs of change here too. Among secular Jews in Europe and America, eating pork, usually as ham or sausages, is not uncommon. In Israel, pork is not officially eaten, but it is known by the name of white meat (*basar lavan*) or by the name of 'other meat' (*basar acher*). This is a source of irritation for strict believers. The demand for pork increased in the 1990s due to the large influx of non-religious immigrants from the former Soviet Union. Many of these migrants missed pork products that were available in the former USSR. Moreover, there were enough entrepreneurs among the newcomers to meet the demand for pork. The number of outlets increased, whereupon local governments took measures to restrict or ban sales. As a result, sales were driven out to remote places that people without cars found difficult or impossible to access. But in 2004, the Israeli Supreme Court ruled that buying and selling pork was no longer prohibited, and that these local bans were not legal.

The debate about eating pork in the Jewish tradition was further fuelled in late 2009 by the publication of a pork cookbook. The title of the cookbook translated from Hebrew is *The White Book*, referring to the term 'white meat'. It was written by a well-known cardiologist, Eli Landau (1949–2012), who used pork from a Galilee farm. Landau was targetting the younger generation of Jews who aspired to a more international style of living. In an interview in the *New York Times* (29 September 2010), Landau said he expected eating pork to be commonplace within 20 to 30 years; however, he didn't live long enough to find out.

Ritual slaughter

In Islamic and Jewish traditions, animals must be ritually slaughtered, otherwise the meat is unclean and therefore inedible. For Muslims, meat must be *halal* (Arabic: clean). As in the Jewish dietary laws, the Quran does not allow the eating of animals that are already dead (cadavers) or the consumption of blood. Muslim butchers slaughter large animals by cutting the carotid artery with a sharp knife. For smaller animals, they cut the throat. Before slaughtering the animal, the butcher utters the following words: 'In the name of Allah. Allah is great.'

In the second half of the nineteenth century, the upper middle class in the Netherlands began to think differently about animals. Popular entertainment like cat-in-the-bag clubbing, eel pulling and goose pulling were increasingly perceived as cruel and, above all, uncivilised. Local animal protection associations emerged to campaign against unnecessary animal suffering. However, as veterinary historian Peter Koolmees describes in his dissertation on municipal slaughterhouses, before 1870 Dutch people were not that concerned about slaughter methods. This began to change with the advent of animal protection and with the further professionalisation of veterinarians at the end of the nineteenth century. The central question was twofold: to what extent was ritual slaughter permissible if it involved animal suffering, and would a ban violate freedom of religion? This led to a fierce debate in which opponents weren't afraid to use terms like 'cruel' and 'repulsive'. The Meat Inspection Act of 1913 was amended in 1920 with an obligation to stun animals before killing, Koolmees writes, but an exception was made for Jewish ritual slaughter. This requires cutting the soft tissue parts of the animal's neck in one continuous up-and-down motion. Bleeding out is essential. To check the correct application of these rules, Jewish butchers are under rabbinical supervision.

In the 1960s, the arrival of Muslim Turkish and Moroccan migrants rekindled a debate on ritual slaughter in the Netherlands. Although the demand for *halal* meat grew rapidly, Islamic slaughter was not (yet) allowed. So that they could nevertheless obtain *halal meat*, some Muslims resorted to slaughtering sheep themselves in their backyard. From a public health perspective, this was seen as an undesirable development. Despite protests from animal welfare organisations, Islamic slaughter was therefore included in the law. The debate then flared up again in 2011 when, again, the question was raised whether limiting animal suffering outweighed freedom of religion. On 28 June 2011, the House of Representatives adopted a bill by the Party for the Animals against unanaesthetised ritual slaughter. Although this bill was eventually rejected by the Senate, the rules around ritual slaughter in the Netherlands were tightened in 2016. In 2018, the legal rules on ritual slaughter varied across European countries (Figure 27). This may change, as in 2020 the European Court of Justice ruled that a Belgian ban on unanaesthetised slaughter does not violate European law or freedom of religion.

FIGURE 27   Legal rules for ritual slaughter in Europe in 2018. Only in Slovenia was
ritual slaughter prohibited. Other countries did not require stunning
(red) or required stunning at different times during slaughter: after
(light green), at the same time as (blue), before (dark green) cutting
the throat.
SOURCES: LIBRARY OF CONGRESS, LEGAL RESTRICTIONS
ON RELIGIOUS SLAUGHTER IN EUROPE (2018) AND SILVIO
FERRARI AND ROSSELLA BOTTONI, LEGISLATION ON RELIGIOUS
SLAUGHTER. FACTSHEET, DIALREL, WIKI COMMONS

The sacred cow

Hindus from India travelling in Europe or the United States are horrified when they
see a cow cut up in pieces and packaged on supermarket shelves. India's sacred
cow is the zebu (*Bos taurus indicus*), a humpbacked cow descended from the *Bos
nomadicus*. It is the most common cattle in both India and Africa. Views on cattle
within Hinduism go much further than among African pastoral peoples: the cow
should in fact not be eaten under any circumstances.

In Africa, animals with which the clan has a close relationship are not eaten. The pastoral peoples of West and East Africa eat beef only occasionally, because the animal actually has too special a place in the culture. The larger the herd, the greater the prestige of the owner and his clan. The cow is usually sold only out of necessity when money is needed. They are bought by farmers, who do eat beef. They also buy the milk and dairy products prepared by women and sold in the market. Today, the pastoral people are increasingly being forced into permanent residence. And due to overgrazing and armed conflict, they are more and more marginalised and lead an existence with few prospects.

Buddhism discourages the eating of four-legged animals in general, and thus also beef. Incidentally, according to *Theravada*, the original Buddhist teachings and traditions, eating meat is an ethically neutral act that cannot cause bad *karma*. After all, when eating meat, the intention is not to kill a living animal. For monks, however, this does mean that slaughtering animals is a wrongful act. In largely Buddhist Japan, eating meat was forbidden for a long time. It was only in 1872, shortly after the Meiji Restoration, that this ban was finally lifted.

Hinduism

The slaughter ban and taboo on beef was not always a central aspect of Hinduism. Beef was eaten in the Vedic culture, a cattle-breeding population that dominated northern India between 1800 and 800 BCE. These Indo-Aryans were originally pastoral peoples. During religious events, priests and war chiefs slaughtered cows and the meat was distributed. All this was done under the supervision of Brahmin priests. Increasing population pressure created a tension between the space needed for grain production and animal husbandry. In this context, appreciation for beef eventually disappeared. The Hindu ban on beef came about because of and was a central tenet of religion. Western visitors in countries like India and Nepal are sometimes surprised by the presence of cows in the streets (Figure 28). They find the idea of the sacred cow slightly amusing. However, the special place of the cow in religion and Indian society can be easily understood if we look at its various functions (see box).

The ban on the slaughter of cows and eating beef in India has gained new momentum with the advent of Hindu nationalism. These nationalists oppose the influences of globalisation, which in their view is at odds with ancient Hindu values. This is happening much to the dismay of India's Christian and Muslim minorities. The sensitivities around beef were highlighted when in 2001, two Americans of Indian origin sued McDonald's in America for using beef fat for frying French fries. Following this, in New Delhi, a fundamentalist Hindu group vandalised a McDonald's restaurant. Subsequently, the Indian headquarters of McDonald's stressed that, as agreed in India, they always use vegetable products for frying French fries and other products. Like McDonald's products, pizza and other non-Indian dishes and cuisines have also become very popular among the Indian middle class, which is also a source of irritation among Hindu fundamentalists.

FIGURE 28   The sacred cow on the street in Kathmandu, Nepal
PHOTO BY FRANCISCO ANZOLA, WIKI COMMONS

Significance of the cow in Hindu culture

– Supplies dairy: milk, ghee and curd.
– Supplies manure. Manure is used as fuel for preparing food, for fertilising
  fields and sometimes for making floors hard and smooth.
– Supplies urine. Urine is important for symbolic cleansing, e.g. of the kitchen,
  a place that is highly susceptible to religious contamination according to
  Hindu culture.
– Works as a draught animal in agriculture.
– Supplies meat and hides. Beef is eaten in India by the untouchables caste
  (*Dalits*) and by religious minorities such as Muslims and Christians.

The beef issue is high on the political agenda. In Kerala, a state in the far south-west
of the country, and in the north-eastern states, cow slaughter and the sale of beef is
still allowed, but there have been attempts to ban the sale of beef throughout India.
For instance, the Hindu nationalist BJP party wanted a total ban on beef during
the October 2010 Commonwealth Games in India. Moreover, at the Copenhagen

climate summit in November 2009, India's environment minister suggested to Western counterparts that they follow the Indian example and stop eating beef. According to the minister, this would lead to a significant reduction in CO_2 emissions. A proposal with a large grain of truth, but one that will be difficult to achieve because it interferes with ancient and deep-rooted food cultures of which beef is an essential part. As yet, the situation has not improved. The slaughter ban in some parts of India has led to a nuisance of stray cows and an opaque slaughter industry with ten thousand illegal abattoirs and processing plants.[1]

A horse is a noble animal

Horse meat was and is a well-known food among Central Asian peoples and also in Japan. Horse meat from young, fat mares, for instance, is highly valued by the Kazakhs (the Qazaqtar), a Turkic people in Kazakhstan and neighbouring areas. They keep horses for consumption. In Europe, the eating of horse meat was common until the eighth century, also in England. The pagan Celtic and Germanic peoples of western Europe used to sacrifice a horse during ritual feasts, after which the meat was eaten. Archaeological finds from the Gallic-Roman city of Lutèce, today's Paris, suggest that horse meat as well as dog meat were eaten from time to time.

Horse meat banned

Although the Greeks and Romans already abhorred the idea of eating horse meat, it was also considered pagan by the Christians. When Boniface was commissioned by Pope Gregory III (731–741) to Christianise the Frisians in 732, he was also told that the rituals surrounding horses and eating horse meat should be banned. Gregory III's successor, Pope Zacharias I (741–752), reaffirmed the ban on eating horse meat, which was the main reason why this food item disappeared from the diet of Europeans in the Middle Ages. The religious significance of the ban was gradually forgotten and, until the nineteenth century, people were convinced that horse meat was harmful to eat.

Back in the nineteenth century

An important impetus for the change in the negative attitude towards horse meat was provided by the many Napoleonic wars and later the siege of Paris during the Franco-German War (1870–1871). In Napoleon's time, soldiers and civilians were so hungry that they ate the horses killed on the battlefield. In besieged Paris, horses were slaughtered out of necessity. The horse meat turned out to be neither poisonous nor dangerous—in fact it was even quite tasty. French pharmacists, hygienists and vets then promoted horse meat as cheap meat for the working population in the cities. An ordinance issued by the prefect of police in Paris in 1866 allowed

---

1 'The sacred cow as a Hindu nationalist weapon', *de Kanttekening,* 31 July 2020.

FIGURE 29
Horse butcher Jakob Levie Wallage in
Stadskanaal poses in front of his shop with his
family, 1920–1930
FROM THE IMAGE DATABASE OF GRONINGER
ARCHIVES

the sale of horse meat without too many problems. However, there were specially
designated horse butchers, who had to be clearly distinguished from the ordinary
butcher to avoid confusion and fraud.

Interestingly, with the rise of railways in the mid-nineteenth century, the consumption of horse meat increased. Carriages with horses were used for local transport of passengers to and from the railway station. This eventually created a large
supply of discarded horses in the cities. A major advantage of meat from old horses
is that it is still relatively tender compared to, for example, meat from an old cow.
Horse butchers appeared in all major cities of France, Belgium and the Netherlands,
offering cheap meat (Figure 29).

Good and cheap

Although the ban on horse meat disappeared, it remained a food of low social status. Nonetheless, horse steak tasted good to many Europeans, and it was a lot more
affordable than beef steak. Smoked and salted horse meat was also cheaper than
its beef equivalent. In the Netherlands, consumption of horse meat increased after
1870. To meet the demand, discarded mine horses were imported from England.
According to Peter Koolmees, horse meat accounted for about nine percent of total
meat consumption in Amsterdam at the beginning of World War I. After that, consumption dropped to four percent in the 1920s. At the end of the 1920s, following
the example of the United States, the phenomenon of the *radio chef* appeared. A
very well-known radio personality at the time was P.J. Kers jr., who gave cooking

FIGURE 30   Horse meat sausages at the market in Tampere, Finland (2015)
PHOTO TAKEN BY NADJA VAHLQVIST, WIKI COMMONS

shows, first for the AVRO and later for the VARA. When asked in one of his cookbooks (1937) about which meat represented the best nutritional value for one guil-
der, his answer was unequivocal: horse meat and mutton.

The extent to which horse meat is eaten varies widely between European countries. In England, for example, it is not popular at all. There, it is considered abhorrent to eat such a 'noble' animal. In Italy, France and Belgium, horse meat is eaten,
but consumption is decreasing. Only 8.1 percent of French households say they buy
horse meat, compared to 11.8 percent in 2016. The number of horses slaughtered for
food fell 15 percent between 2019 and 2020 to less than 6,840, the lowest number in
a decade, three times less than in 2013.[2] In the Netherlands, there are only two to
three thousand because they are not allowed to be bred for slaughter. Most horse
meat is imported from Argentina. Horse steak and smoked and salted horse meat
are popular, but also snacks and sausages (Figure 30).

Vegetarianism and veganism

Throughout evolution, humans have always eaten animal products. However, vegetarianism as we know it in the Netherlands dates back to the nineteenth century.

---

2   M. Witdouck, 'Consumption of horsemeat falls sharply in France', *Food & Meat*.

Following the movement in England, the Dutch Vegetarians' Union was founded in 1894. The Bond published a number of cookbooks to make it easy for people to prepare a tasty meatless meal. Other organisations also sometimes published vegetarian recipes (Figure 31). The first vegetarians were mostly women and men from more affluent circles. It wasn't until almost a century later, in the 1970s, that vegetarianism began to gain more ground, partly due to the rise of social groups expressing concerns about animal welfare and the environment. In our age of cheap meat, a diet without meat, once a sign of poverty, can in fact signal an increase in status.

Interestingly, vegetarian movements are mostly active in the northern part of Europe. In southern Europe, vegetarianism never got off to a good start. Perhaps this has to do with a moralising Protestant ethic and culture of right and wrong in northern Europe. There are more examples of these European differences. The alcohol restrictions introduced by governments in northern Europe in the late 19th century, for example, was not really emulated in southern Europe either.

While vegetarians specifically do not eat meat, vegans are much more strict. Veganism is a way of life that does not involve using animals. The problem with a dairy-free vegan diet, which does not contain products of animal origin, is that it does not contain an adequate supply of protein, iron, vitamin B1, vitamin B2, vitamin B12 and calcium. Today though, thanks to modern food science and the pharmaceutical industry, it is possible to supplement deficiencies. The *Vegan Disc* provides information on a balanced vegan diet (Figure 32), listing for example certain fruits and vegetables that are a source of calcium. Furthermore, vegans are advised to take vitamin D and B12 and omega-3 supplements.

In the Netherlands and other Western countries, there is a growing debate about whether killing and eating animals is ethical. Dutch environmental and animal

Vegetarisch gehakt.
(om koud op de boterham te eten).

100 G. pinda's (of meer).
3 beschuiten.
3 sneedjes oud bruin brood.
3 koude aardappelen.
1 ui, een beetje kerri, zout.
1/2 theelepeltje kokskruiden.
wat peterselie en selderyblaadjes.
1 ei, iets water of melk.
paneermeel, boter.

Bereiding. De geroosterde pinda's fijn malen, ook beschuiten, brood en aardappelen, de ui (of een paar sjalotten), de peterselie- en selderyblaadjes fijn hakken. Dan alles dooreenmengen, tot het gelijkmatig vermengd is, daarna het ei er in breken en wat zout. Doorgaans is het mengsel hierdoor niet vochtig genoeg; er moet dan een beetje water of melk toegevoegd worden, ook een beetje gesmolten boter, als men het lekkerder wil maken. Nu op de gewone manier een bal er van maken, wat boter bruin braden, de bal er in leggen, met paneermeel bestrooien en ongeveer 3/4 u. in den oven braden. Voor het aansnijden geheel koud laten worden.

FIGURE 31
A 1920 recipe for 'vegetarian mince', from the Dutch magazine 'In en om de woning, het tijdschrift van de huishoudscholen'

FIGURE 32   The vegan disc

rights organisations like *Wakker Dier, Varkens in Nood* and *Milieudefensie* adver-
tise in the media and publish informative material to show what is wrong in the
meat sector. This makes it harder for many people to eat a meat-based meal without feeling uncomfortable. Consumers may wonder whether the meat has been
produced in an animal- and environment-friendly way, or what the consequences
are for farmers in the Netherlands and in developing countries. It has become a
global issue.

The debate was further fuelled in April 2010 by a plea from over 100 professors for sustainable livestock farming. Their action received a lot of media attention. On the website, they wrote: 'The sector has become ever more intensive and
large-scale; healthy animals are still being culled on a massive scale, animal disease crises are a threat to public health, and the living and slaughter conditions of

hundreds of millions of animals in our "civilized" country are still shameful.'[3] As a tool for consumers, the Meat Guide (*VleesWijzer*)was introduced in 2009. With this, the Varkens in Nood Foundation, together with *Milieudefensie*, informs consumers about animal-friendly produced meat. Clearly, whether we like it or not , there are moral issues on our plate.

A similar debate is ongoing in the United States. Back in 2001, journalist and writer Eric Schlosser was already taking a critical look at cheap meat and the bio-industry in his notorious book *Fast food nation*. A few years later, the writer Jonathan Safran Foer published his book *Eating animals*, the Dutch translation of which appeared in 2010 (and a documentary, *Eating animals*, in 2018). American psychologist Melanie Joy, author of the bestseller *Why we love dogs, eat pigs and wear cows: an introduction to carnism*, of which a Dutch translation was published in 2021, introduced the concept of carnism as a counterpart to vegetarianism. She regards the cultural habit of eating meat as a learned habit with ideological traits.[4]

Meat substitutes prove to be no substitute (*yet*)

Research on protein substitutes shows that meat eaters often have a fear of new, unfamiliar meat substitutes, a 'food neophobia'. This could be a barrier to the acceptance of meat substitutes. Work on meat substitutes made with plant products, in particular soy proteins, has been going on for a long time. In the Netherlands, in Wageningen and Utrecht, research is being done on meat substitutes that resemble meat. According to Wageningen research, meat substitutes should resemble meat sensorially, externally and in terms of taste. Combinations of vegetable and animal proteins are also being considered.

Plant-based meat substitutes such as tofu, pulses and vegetable burgers are a full-fledged alternative to meat, and they have much less impact on the climate. Meat substitutes are gaining in popularity, with production and supply growing every year. Yet meat substitutes do not yet appear to live up to the 'substitute' promise. According to figures from 2020, meat substitutes have a market share of only 2.5 percent. Dutch people spend about 17 euros on them per year and eat 870 grams of these items. Yet this means the Dutch eat the most meat substitutes within Europe.[5]

More on meat

In 2001, *De Volkskrant* published an article on the temptations of eating meat. 'Meat is afflicted with a "devilish ambiguity",' Adel den Hartog said in this piece.[1] By this he meant that in all cultures people realise that they have to shed blood to eat meat, have to kill to eat. Meat is sinful. Thinker of the Fatherland, René ten Bos, calls eating meat 'totally immoral'.[2] Clearly, meat consumption in the Netherlands is an ethical issue.

---

3   Full plea on Sustainable livestock production.
4   https://carnism.org.
5   Verena Verhoeven, 'Dutch eat most meat substitutes in Europe,' *Het Parool*, 10 May 2021.

Three reasons are usually given for eating less or no meat: animal welfare, human health impact, and environmental and climate impacts. Health effects have been identified relatively recently. Eating a lot of processed meat and red meat, for example, has been found to be associated with a higher risk of stroke, type 2 diabetes, colon cancer and lung cancer.[3] Concerns about climate change reinforce vegetarians' environmental argument. In a recent *Voedingscentrum* poll, four in 10 Dutch people (41 percent) say that eating meat every day is now a thing of the past. Environmental impact (37 percent), animal welfare (34 percent) and health (27 percent) are reasons for them to cut back.[4]

Growth of global meat consumption

Globally, meat consumption is nevertheless on the rise. While meat used to be a luxury, today eating meat has been democratised: it is cheap and accessible to almost everyone. Based on meat consumption figures, we can see that meat is popular. Den Hartog predicted in 2001 that Western countries would start eating less meat, while developing countries would start catching up big time. The latter has come true, but there has been no structural decline in the West.

In 1995, 198 million tons of meat were eaten worldwide.[5] By 2018, meat consumption was already at 360 million tons; more than half of that increase can be attributed to population growth and 46 percent to higher per capita consumption. Figure 33 shows the growth in meat production. There are large national differences in meat consumption. The Netherlands is in the higher regions with 77.8 kilos per person per year in 2019.[6] But the US and Australia are even bigger meat eaters, consuming more than 100 kilos per year.[7] By contrast, people in countries like Ethiopia consume only 4.2 kilos of meat per person per year.[8] Incidentally, this is total consumption in carcass weight;

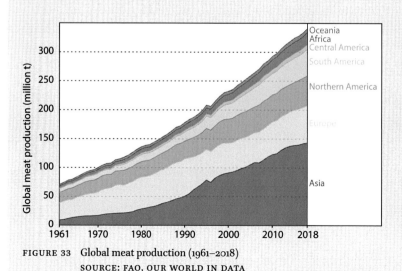

FIGURE 33   Global meat production (1961–2018)
            SOURCE: FAO, OUR WORLD IN DATA

actual meat consumption (without bones) is about half that, which in the Netherlands amounts to 39 kilos per person per year.

No decline in meat consumption in the Netherlands

After 2010, meat consumption in the Netherlands seemed to start declining, but this trend has not continued. Since 2016, more meat—especially chicken—has been eaten again. The Disc of Five recommends no more than 500 grams of meat per week, which amounts to a total net meat consumption of 26 kilos, versus the current 39 kilos per person per year. Interestingly, meat consumption is rising slightly, while at the same time the number of people saying they eat less meat is increasing. Figure 34 shows what people say about their own consumption behaviour.

Since the introduction in 2009 of the Meat Guide, animal welfare and environmental considerations have been regular features of the discourse on

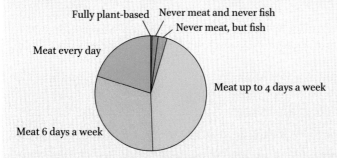

FIGURE 34   Meat and fish consumption in the Netherlands 2020 among
people aged 18 years and older
SOURCE: STATISTICS NETHERLANDS, 'MEAT NOT A DAILY
MEAL FOR 8 IN 10 DUTCH PEOPLE', 9 JUNE 2021

FIGURE 35   Superlist Green 2021
SOURCE: QUESTIONMARK, PHOTO DENIZ ÜLBEGI. REPRODUCED WITH
PERMISSION

meat. While the Meat Guide no longer exists, the data can be found at *Questionmark*. This organisation is committed to a market 'whose focus is healthy, sustainable and ethically responsible'.[9] The organisation claims that, in practice, little is yet visible of the agreements made by supermarkets on promoting sustainable food.[10] According to Questionmark's *Superlist Green 2021*, supermarkets are not yet making a substantial contribution to reducing meat consumption (Figure 35).

Environmental impact

Livestock production is a major contributor to the production of greenhouse gases. Meat has such a large climate impact because an average of five kilos of plant-based feed is needed to produce one kilogram of meat. Environmental organisations and scientists therefore point out that eating less meat is good for the climate.[11] However, there are no measures in the 2019 Dutch Climate Agreement to ensure that people eat less meat. Parties hope that with incentive measures people will make a sustainable choice themselves. That seems like a vain hope. According to a UN projection, global meat consumption will increase by 76 percent by the middle of the 21st century. What impact will this have on the environment?

Conclusion

Meat consumption in the Netherlands cannot be expected to change quickly by itself. The prerequisite is that institutions succeed in bringing about a change in norms and values. It is therefore essential for authorities in health care, government and social institutions to have a good understanding of the place of meat in our society. Wageningen researcher Hans Dagevos reiterates what Adel de Hartog also said at the time: 'Meat appears to be strongly embedded in our food culture'.[12]

Sources

1   Mac van Dinther, 'Begeerlijk vlees', *de Volkskrant*, 27 March 2001.
2   Henk Steenhuis, 'Waarom het eten van vlees "totaal immoreel" is', *Trouw*, 7 March 2018.
3   Voedingscentrum, 'Waarom is minder vlees eten beter voor je gezondheid en het milieu?'.
4   Voedingscentrum, 'Nederlanders eten minder vlees voor milieu en dierenwelzijn', 20 July 2021.
5   Van Dinther, 'Begeerlijk vlees'.
6   'Nederlander at alweer meer vlees in 2019', *Foodlog*, 22 October 2020.
7   David Widmar, 'U.S. Meat Consumption Trends and COVID-19', *Agricultural Economic Insights*, 5 April 2021.
8   Data from 2013. OECD, 'Meat Consumption'.
9   Questionmark, 'Over ons'.
10 Questionmark, 'Hoe dragen supermarkten bij aan duurzaam voedsel?'.
11 See for example: Milieu Centraal, 'Vlees'; H. Charles J. Godfray, *et al.*, 'Meat consumption, health, and the environment', *Science* 361 (2018), eaam5324.
12 NOS, 'Opnieuw meer vlees gegeten in Nederland', 22 October 2020.

CHAPTER 5

Rats, dogs and insects—delicious or disgusting?

In the Netherlands, meat is mainly eaten from animals bred for meat consumption—pigs, cattle, sheep, chickens and other poultry. Game, which is in supermarkets especially around Christmas, is often not wild at all, but bred for consumption. The same goes for many fish, such as salmon, which is reared on a large scale in fish farms. But we are also able to eat pets like Flopsy the domestic rabbit. Since very many animals can in fact be eaten, the choice of which animal is considered edible is quite arbitrary.

Rats and rodents

During the Biafra War in Nigeria (1967–1970), reports circulated in the Netherlands about starving people eating rats. Famine was indeed rife in Biafra province, which lies in and around the oil-rich Niger Delta region. But rats and other rodents had been part of the local food culture for centuries. Western journalists—prejudiced against rat meat as food—picked the wrong symbol to highlight human misery.

Rats have a bad reputation in Europe: not only are they vociferous eaters that feast on food crops and supplies, but they are also associated with war, hunger and the transmission of dangerous diseases. In particular, the transmission of plague by rats largely explains Europe's aversion to these animals.

Rodents are eaten all over the world, but it is difficult to say in what quantity. In any case, the order of rodents, *Rodentia*, includes as many as 2,277 species, among them rats, mice, porcupines, guinea pigs, squirrels, beavers and hamsters. Rabbits and hares belong to another order, *Lagomorpha*, and are a popular food in the Western world.

In Europe, squirrels are hunted on a modest scale for their meat in France, Italy and Greece. Another exception is the European edible dormouse (*Glis glis*) from the dormouse family. Wild dormice are still caught in Slovenia and eaten as a delicacy. And back in the day, the Romans considered the dormouse a tasty snack. The animals were fattened and cooked in special terracotta pots. A complicated preparation of dormouse can be found in the Roman cookbook *De re coquinaria* from the fourth century.

Europeans otherwise ate rats and mice only in times of war and famine. Famous examples are the siege of Paris in the Franco-German War (1870–1871) and the siege of Leningrad (St Petersburg) during World War II (1941–1944).

Rodents are commonly eaten in countries with few large farm animals such as cattle, pigs, goats and sheep. For example, the guinea pig (*Cavia porcellus*) has been bred and eaten in the Andes in South America since the time of the Incas. The Himalayan marmot (*Marmota himalayana*) is eaten in Tibet and other parts of the Himalayas. The Aborigines of Australia eat a variety of rodents, including rats. In Southeast Asia too, rats are eaten, including in Thailand (Figure 36). Rising inflation has greatly increased the demand for more affordable rat meat in Cambodia. Live rats are therefore exported to Cambodia from Vietnam for consumption. In Oceania, eating rat meat is common practice. For example, in French Polynesia, in the Society Islands, catching and eating rats is an old custom. In these places, caution should be exercised when using pesticides: like insecticide, rat poison can be dangerous to humans if the meat is part of the food culture.

Bushmeat

In tropical Africa, to the bewilderment of Europeans, eating rodents is an ancient phenomenon. My first experiences with these aspects of African food culture date back to the early 1970s in Ghana. Along roads in the south of the country, vendors approached us with large rat-like animals which they loudly touted as *bushmeat*, to be translated as 'game'. It was the *greater cane rat* or *grasscutter* (*Thryonomys swinderianus*)—a highly sought-after piece of meat—and it tasted good!

In the savannahs of Africa, *gerbils* (desert rat or racoon), which are common there, are eaten. In nutrition research, food obtained from hunting is quite often placed under the English collective term *bushmeat* or the French *viande de brousse* or *gibier*. This makes it difficult to ascertain exactly which rodents are eaten. Moreover, it is difficult to find out which small animals children collect in the fields. Boys go out with their slingshots to shoot birds and set snares to catch small rodents.

In Zimbabwe, they use a different technique: after burning down the grass savannas in September and October, the mice and rats that escape the fire are caught and then eaten. In a good season, it's possible to catch about two hundred mice; sometimes for weeks on end an average of one mouse a day. In Lusaka, Zambia's capital, rats and mice are sold in large quantities in the market. Here it is not a delicacy like the *greater cane rat*.

It's important to say that not everyone in tropical Africa eats rodents. Muslims, for instance, are not permitted to eat carrion rodents, including mice and rats, and you will seldom come across this practice in this area. This is evident among the pastoral peoples, such as the Fulani or Peul of the West African savannas, as I described in a 1974 article in the journal *Voeding*.

Rat meat and other rodents are an underexplored source of animal protein. A growing problem for populations on the African continent is that it is becoming less and less profitable to hunt and trap small and large game in fields and forests for their own consumption. Rats and mice have high reproductive potential, but the situation is different for other rodents and larger game. Population growth, deforestation and commercial hunting threaten the sustainability of game. Outside game reserves, the availability of *bushmeat* in West and Central Africa has dramatically

FIGURE 36   Roasted rats for sale at a roadside stall near Suphanburi in Thailand
         PHOTO BY USER GRENDELKHAN, WIKI COMMONS

reduced, be it *grasscutters*, *pangolins* (*Mandidae*), antelopes, porcupines, wild pigs or monkeys. As a result, *bushmeat* hunting as a food source is in crisis. For the rural population in the Congo Basin, *bushmeat* is still an accessible but also a declining source of income due to sales to the city, where it an expensive item on the menu in restaurants. It is prestige food for the urban elite who see this as a means of maintaining their emotional connection with the countryside. Conservation organisations, in consultation with local people and authorities, are trying to develop *wildlife farming* programmes for the more sustainable management of game as a food source and for maintaining biodiversity. In 1999, the Bushmeat Crisis Task Force, a consortium of wildlife organisations and scientists, was set up for this purpose.

Dog as a delicacy?!

Animals, especially small pets, are usually not eaten because people have a special emotional bond with them. It is the main reason why dogs and cats are not eaten in Western societies. In our urban society, it is inconceivable, even terrible, that people would slaughter and then eat their own pets. And anyway, the abundance of other types of meat makes it unnecessary. Exceptions can be found during periods of great food shortages. During the aforementioned siege of Paris (1870–1871), French city dwellers ate not only rats, but also dogs and cats. During the Dutch famine of 1944–1945, dogs and cats were occasionally eaten in the western part of the Netherlands, often without people realising. Euphemistically, an edible cat was called a 'roof rabbit'. Even in this crisis situation, the emotional connection with these animals was too strong for them to be widely accepted as emergency food.

In the literature we can see that cat meat in particular was (and is) historically eaten very rarely. In ancient Egypt, the cat was a sacred animal and was often mummified. Medicinal properties were attributed to body parts and excretions of the cat. As far as dog meat is concerned, the situation is different. There are benefits to be had from eating dog meat. It is protein-rich, for one thing. Moreover, in densely populated areas, dogs, like pigs and chickens, can as stray animals largely scratch together their own food. These scavengers also need much less space than, say, cattle. So the dog has some similarities with the pig. During its domestication in the Middle East (12000 BCE), the dog probably had numerous functions among the farming peoples and shepherds, such as guard dog, playmate, herder of the roaming cattle, helper with hunting, but also as a source of food. In Mexico among the Aztecs and in Central America, it was more common to eat dog meat. Furthermore, in classical antiquity, dog meat was a familiar item on the menu among the Greeks and Romans.

Even now, dog meat is eaten in various places. It is common in West Africa and especially in the savannas, although not in all populations. I came across it during my travels in parts of East Asia, including China, the Philippines, Vietnam and Korea (Figure 37). In North Vietnam, dog meat is very popular and is sold in markets and shops. Vietnamese, incidentally, are aware that Westerners look at it with horror. When I took photos of a stall with dog meat on display in a market in Hanoi, I received furious reactions. People thought I was a Western meddler, asking 'what was I doing there?'.

China's meat supply has traditionally relied on a trio of animal 'lowlifes': pigs, chickens and dogs. In China, dog meat has been eaten for millennia. It is also used there as medicine, as it is believed to have aphrodisiac properties. Dogs and cats are bred specifically for meat and are highly sought after by the rural population. Among others, people breed the chow chow, one of the oldest Chinese breeds, to serve as a hunting dog and watchman, as well as for eating. In recent years, with the rise in prosperity, keeping dogs and cats as pets is becoming increasingly popular among the Chinese middle and upper classes. As a result, the Chinese Association for the Protection of Small Pets is now trying to ban the eating of dog and cat meat. There are also plans from the government to ban their sale, which would mean they would disappear from the menu in China. Meanwhile, Shenzhen, located on the border with Hong Kong, became the first Chinese city to ban eating dogs and cats as of 1 May 2020.

FIGURE 37 Adel den Hartog (second from left) on a working visit to the Philippines in the 1980s

A lack of civilisation

In the Western world, it is unthinkable to eat dog meat. Emotionally, many people are so close to dogs that eating them would be an act of cannibalism, similar to eating a close friend. Therefore, in Western society, dogs and cats are not edible. Westerners see eating dog meat as a sign of a lack of civilisation. This was evident during the Louisiana Purchase Exposition, the 1904 international exhibition in St Louis in the United States. This world exhibition featured a Filipino reservation with all kinds of exotic items. A key feature was the rituals surrounding the slaughter and eating of dogs by the Igorot, an ethnic group brought to the exhibition especially from the Philippines. The exhibition highlighted the lack of civilisation and thus the need for an American civilisation offensive in the Philippines, which had been conquered from Spain shortly before. All this, incidentally, happened to the great displeasure of the Filipino elite, who did not want to see their country and culture exhibited in this way. Today's Igorot, who live in the Cordillera mountain range on the Philippine island of Luzon, still eat dog meat. Slaughtering and eating dogs is part of certain ceremonies. For them, the meat is not only a cheap and good source of animal protein, but also has beneficial properties in their view.

Western misunderstanding about dog meat was evident in actions by animal activists at the start of the Seoul Olympics in the summer of 1988, and again at the 2002 World Cup in South Korea (and Japan) (Figure 38). French actress Brigitte

FIGURE 38   A dish of dog meat at a restaurant in Seoul, South Korea. Dog meat tastes a bit like pork.
PHOTO: RHETT SUTPHIN, WIKI COMMONS

Bardot played an important role in these actions. In response, Koreans, with a nice feel for humour, called a soup made from dog meat *Bardot's soup*. On the orders of the Korean authorities, the sale of dog meat and the serving of dog meat stews in restaurants, for example, was kept out of sight of visitors. In this way, they tried to avoid the Western stigma about dog-meat eaters.

A sign that times do change is evident from a recent discussion in South Korea about forbidding the consumption of dog meat. This has a lot to do with the increasing number of South Koreans keeping dogs as pets. According to former minister of justice Choo Mi-ae, 'eating dogs really has no place in a developed country anymore'. Although dog meat is less often on the menu, about one million dogs are still slaughtered every year. According to President Moon Jae-in, a ban would be desirable.[1]

1 Gijs Moes, 'Hondenvlees op het menu is straks misschien verleden tijd in Zuid-Korea'. *Wed*, 28 September 2021.

A dinner in Togo

During a food mission to northern Togo, West Africa, in which I participated, we stayed overnight in a small hotel. The manager had a nice little dog. When we didn't see the animal the next day, I asked where it had gone. The manager kindly told us that they had given the best available meat to the guests. The meat had been incorporated in a sauce. I had not tasted anything special, but it was not a pleasant experience in retrospect. At moments like that, your culturally defined prejudices kick into action.

Insects as food

Some religions have such a reverence for life that they view it as all-encompassing. Jainism, a religious movement in India, for example, is strictly vegetarian. For Jain monks, respect for human and animal life, however small the animal, manifests itself in sweeping the path they walk on to avoid trampling insects. Some monks wear a cloth over their mouths. While some say this is to avoid accidentally swallowing an insect, others say this prevents saliva from falling from the mouth onto a holy book.

Most Dutch people find insects inedible and are horrified at the very thought of consuming them. In contrast, in tropical Africa and large parts of Southeast Asia, people actually find insects tasty. This difference in the popularity of insects as food has a geographical reason. Indeed, in our regions, there are no obviously edible insects. Moreover, larger animals—chicken, sheep, pig and beef—thrive here, and their meat is therefore plentiful. Insects have had no place in food culture in north-western Europe until now, although there is evidence that in times of famine, insect-eating occurred. For example, mayfly larvae were eaten during the siege of Derry/Londonderry in Northern Ireland in 1688, which lasted 105 days and during which half the inhabitants starved to death.

An ancient phenomenon

Eating insects, called entomophagy, is a very ancient phenomenon. Apes are known to eat insects and it is very likely that insects were part of the diet of early humans, Homo sapiens. Among the Aborigines, who settled in Australia 70,000 years ago, insects were always part of their diet.

In ancient Egypt before our era, pictures and papyrus scrolls do not directly show that insects were part of the diet. However, locusts are depicted. Furthermore, there is a papyrus text that mentions the use of insects as a purifying agent or as a magic remedy. The ancient Greeks were certainly familiar with locusts as food. The Greek comedy poet Aristophanes (445–386 BCE) mentioned that they were for sale in the market, and were eaten by the poor. The philosopher Aristotle (384–322 BCE) further recommended cicada larvae as a delicacy. However, according to the Greek philosopher and writer Plutarch (46–120), cicadas were mainly sacred animals that were in fact not eaten much. Pliny the Elder (23/24–79) in his *Naturalis historia* mentioned the use of insects in the world as it was known then, such as locusts among the Parthians in present-day Iran. He also wrote about the consumption of larvae by the Romans. It is possible that these were the larvae of the stag beetle (*Lucanus cervus*). Pliny admired insects, those infinitely finely built little animals. He was well aware of his readers' distaste for many species of insects and advised them to get over it.

Religious food laws regarding insects are sometimes complex. For instance, the Old Testament mentions that some insects may be eaten (see box).

Leviticus 11:22 on eating insects

'Even winged insects are considered inedible to you. You may only eat those that also have a pair of jumping legs. These are the various species of field grasshoppers, sable grasshoppers, crickets and dwarf grasshoppers. You can eat those. All other winged insects are considered inedible to you.'

Source
New Bible Translation

In Islam, a category of animals referred to by a biblical term as creeping creatures is labelled inedible, but locusts are the exception, and they are still considered a delicacy in Egypt and North Africa.

Does unknown remain unloved?

Eating insects is not an illogical thing to do. During the period of colonisation of Africa and Southeast Asia, Europeans often condescended to the habit of eating insects. In East Africa, the mission taught in schools that eating insects was an uncivilised custom. Europeans did not disguise their distaste. Eating insects did not disappear as a result, but people were increasingly reluctant to talk about it with outsiders.

There have always been prejudices about other peoples' food customs. People from northern China laugh at customs in south-eastern China, where they eat frog legs. The eating of insects by the inhabitants of the southern island of Hainan was considered uncivilised. European settlers in North America abhorred the insect consumption of the indigenous peoples. In the early seventeenth century, settlers on the east coast mistook local lobsters (*New England lobsters*) for insects, and so did not eat them. It took them a while to realise that there were substantial differences and lobster was eventually seen as a delicacy.

Enlightened minds in nineteenth-century Europe thought otherwise, and there are several publications containing positive reports about eating insects. The naturalist Alfred Russell Wallace (1823–1913), one of the founders of biogeography, observed while travelling in Southeast Asia that the inhabitants took advantage of the abundance of insects by simply eating them. A booklet by Vincent Holt, entitled *Why not eat insects?*, was published in 1885 and was later reprinted several times. The writer chose the motto: 'Them insects eats up every damned green thing that do grow, and us farmers starves. Well, eat them, and grow fat.' He pleaded for Europeans to eat and appreciate this underexposed food. Holt asks readers to follow his arguments honestly and impartially, pointing out that insects are herbivores, clean, perfectly edible and nutritious. Insects, according to Holt, are eaten by both 'civilised' and 'uncivilised' peoples. The physician Cornelis L. van der Burg (1840–1905), in his 1904 book on the diet of the Dutch East Indies, also gave an account of the most commonly eaten insects including white ants, bee larvae, grasshoppers and cicadas. According to him, roasted insects taste like almonds.

More recently, authors have tried to popularise insects as food. The authors of *The Malawi cookbook* (1979) rightly argued that insects are not only delicious, but also a good and cheap source of protein and fat. Western visitors should try to put aside their disgust, they believed, because taxonomically insects are not that far removed from shrimps—which are a valued food in the food culture of Europe and North America.

Where and how much?

The place of insects in the diet is mainly determined by geographical conditions and the prevailing agricultural system. In areas with a relatively high availability of meat, dairy and fish, insect consumption does not make much sense. Insect consumption occurs where the environment is rich in insects but at the same time poor in farm animals. Insects have a place in the diets of tropical Africa, the Amazon, Mexico and Central America, and Southeast Asia. Among the Aborigines of Australia, insects are still an important element of food culture.

In countries such as Mexico and Thailand, insects are sold not only in the market (Figure 39), but also as preserves in cans or bottles, or dried in plastic bags. The number of edible insect species in the world has been estimated at 1,400. How many of these occur in a particular region can vary greatly: in Mexico, for example, it is about 350 species, in the Central African Republic 180 and in Japan 120. The main edible insect species are: caterpillars, silkworms, locusts, beetle larvae, pupae of bees and wasps, bugs, termites, ants and mosquitoes (including water mosquitoes from the East African lakes).

It is difficult to determine how many people eat insects. In tropical Africa, for example, research on insect consumption is hampered by a certain embarrassment the rural population has developed over time about eating insects. When talking to urban researchers and other strangers, people are cautious about discussing insects as food for fear of being perceived as uncivilised. Another problem is the seasonal

FIGURE 39   Fried insects at the market in Thailand
PHOTO BY USER JNPET (2017), WIKI COMMONS

availability of insects. There is an abundance of insects in the rainy season, but that decreases when it gets drier. This may sometimes have led to an underestimation of consumption. There are extensive lists of names of edible insects, but data on exactly how many people eat them are still limited. One exception is a French study among the Oto, a people living around Lake Tumba in Congo, which shows that consumption of caterpillars ranges from 2–4 grams per head per day. For them, this is a modest amount compared to other foods of animal origin.

As mentioned, the nutritional value of insects is good. They contain high levels of high-quality protein and a reasonable amount of fat. In the Western world, finding sources of protein and fat is not much of a problem. Here, about 38 percent of total dietary energy comes from fat, a figure that nutrition scientists say is far too high and has been linked to the incidence of cardiovascular disease. However, protein and energy malnutrition are still common in several poor countries. In some developing countries, such as Malawi, the share of dietary energy derived from fat is only around 10 percent. Young children there cannot process large quantities of porridge or gruel that contain only a little fat to extract enough nutritional energy. Children's stomachs are simply too small for such large amounts of starch. On top of this, extremely low-fat diets often come at the expense of taste.

Interest in the Western world
In the late 20th century, there was renewed interest in Western countries in insects as food. Ronald L. Taylor, professor at the University of Southern California, raised the issue with his book *Butterflies in my stomach: insects in human nutrition* (1975).

The work received considerable attention, but had no practical effect on possible improvements to the world food problem. In 1989, an equally high-profile book by Swiss biochemist Bruno Comby entitled *Délicieux insectes. Les protéines du futur*, on insects as the source of protein in the future. This book was also well received: many found the consumption of insects a strange but interesting idea. Comby pointed out that insects can convert plant material into protein and fat very efficiently (as much as 20 times more efficiently than cattle), but again Western cultural prejudices about insects proved an obstacle to the proper development of his idea.

Research in the Netherlands

Protein from insects could be significant in meeting a growing global demand for protein. Research on insects as food gained momentum in the late 20th century. Leading the way was the work of Wageningen professor of entomology Arnold van Huis. Research into meat substitutes based on plant proteins such as soy had been going on for some time, but insect protein could also be a good meat substitute.

Farms producing edible insects exist in the Netherlands. One company in Oostzaan produces locusts, buffalo worms and mealworms for human consumption. In cooperation with the Rijn IJssel Vakschool in Wageningen, a number of interesting recipes have been developed for Dutch consumers. However, it remains difficult to overcome the ingrained aversion to insects as food. The most promising technique seems to me to be the addition of insect protein to products such as snacks, ready-made meals and meat substitutes. Perhaps insect protein in meat substitutes could even enhance the meat taste. As long as the insects are processed into an unrecognisable form in the product, the cultural defence mechanism will not play too big a role.

Future projections

What can we expect from insects as food in the coming years? The Ministry of Agriculture made one million euros available to Wageningen University in 2009 for research into the possible use of insects in the food chain. Worldwide, due to population pressure, the contribution of hunter-gathering to the food supply will decrease in rural areas. The contribution of wild plants, and small and large animals as supplements to food will come under further pressure. This could also happen with insects. Moreover, the use of insecticides in agriculture could pose major risks when eating insects. On top of this, societies are increasingly urbanising, and for some, eating insects may not fit into the 'civilised' urban lifestyle. In recent years, however, in countries with a tradition of insect consumption, a re-evaluation has been taking place. There they believe that insects are not only nutritious and tasty, but also part of their own culinary heritage—something worth cherishing in this globalising world.

More on edible insects

The acceptance of insects as food has clearly increased following the release of the FAO publication *Edible insects. Future prospects for food and feed security* in 2013. The report, which has been downloaded seven million times, was prepared under the direction of Wageningen professor of tropical entomology Arnold van Huis. Van Huis and his collaborators consider edible insects to be an important contribution to dietary protein requirements. Moreover, insects can be cultivated in an environmentally friendly and efficient way. Insect farming involves low CO_2 emissions: a kilo of insects produces only one percent of the CO_2 that a kilo of cow meat requires. Insects require less feed, water and space, and they generate less waste. Moreover, their food conversion rate, which is the amount of feed needed to be converted into meat, is much better than that of other animals, as Adel den Hartog points out. A kilo of beef requires ten kilos of feed, a kilo of chicken meat requires four kilos of feed and a kilo of caterpillar meat requires only one and a half to three kilos of feed.

As early as 1999, Wageningen University organised a lecture series 'Insects and society' in which both Arnold van Huis and Adel den Hartog spoke.[1] At the time, Van Huis saw particular potential for insects as a source of food for people in developing countries.[2] Den Hartog's expectations were not high. He believed that on a global scale, insects have a marginal significance as food for humans and only a limited role in solving the global food problem. However, he did see a place for insects as a delicacy in a modern food culture, including in the Netherlands. 'I observe increasing attention to this kind of exotic subject,' he concluded. Ten years later, Den Hartog saw that insects were already becoming more *trendy*. Wageningen University had asked him again to give a lecture at its theme night 'Insects: the meat of the future'. Because food habits are not static, it is possible to promote a trend, Den Hartog explained. On that evening, insects were served. Den Hartog himself was able to appreciate the fried locusts he used to get in West Africa as a kind of bar snack with a beer.[3]

An abiding interest in insects
From 2010 to 2014, Van Huis conducted the four-year research programme 'Sustainable production of insect proteins for human consumption (SUPRO2)'. At the end of the project, the international conference *Insects to feed the world* was organised in 2014 (Figure 40). The aim of the conference was to provide frameworks and a roadmap 'to develop a new food and feed industry to contribute to sustainable global food security'.[4]

Several departments at Wageningen University have subsequently researched whether people are willing to accept insects as food. Results for 2016 and 2017 provide insight into who eats insects and why. It turns out that

FIGURE 40   Insects on the menu
SOURCE: ROB RAMAKER, 'INSECTS ON THE MENU', 16 MAY 2014,
RESOURCE WUR

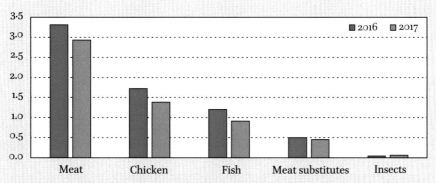

FIGURE 41   Number of times meat, chicken, fish, meat substitutes or insects were eaten in an
average week in 2016 and 2017 in the Netherlands, E-depot WUR

Westerners who eat insects have a certain profile. They tend to be male, young, highly educated, not afraid to try something new and conscious of what they eat. Compared to other protein sources, insects did remain a marginal food (Figure 41).[5]

A difficult market

Several years later, Dutch consumer acceptance of eating insects had not greatly improved. Insects as food still evoked the most disgust in 2019. 'This response generally confirms a consistent finding in consumer studies on edible insects in Western diets,' according to researchers.[6] Eating insects is emotive, that much is clear.[7] However, the European Food Safety Authority (EFSA) recently found the dried yellow mealworm suitable for human consumption. More insects will follow. After all, edible insects fit well with European projects such as the *Green Deal* (for better climate policy) and the farm-to-fork strategy.[8]

Dutch breeders of insect meat have united in the organisation VENIK. Their market is gradually growing. The fact that insects are edible, and are being made so, does not necessarily mean they will be eaten. Jonas House, a researcher in the Consumption and Healthy Lifestyles group at Wageningen University, pointed out the decisive role of the consumer. According to him, going from a network of interested parties to a general embedding of a product in food culture is a huge step.[9]

International literature cites the Netherlands as an example of how insect phobia can be turned around, resulting in acceptance of insects as food. Swedish researchers believe that resistance to certain foods can quickly disappear.[10] But as early as 2010, Adel de Hartog wrote: 'It will take time and effort to get meat eaters to start eating meat substitutes.' It is doubtful that insects will be those meat substitutes.

Sources

1   Hans Kraak, 'Op wereldschaal spelen insecten marginale rol', *Voeding Nu* 1:4 (1999) 20–21.
2   For an overview, see the personal page of Van Huis: Wageningen Universiteit.
3   Liesbeth Koenen, 'Sprinkhaan met bier', *NRC Handelsblad*, 6 January 2009.
4   Wageningen Universiteit, 'Insects to feed the world' conference, 28 June 2013.
5   Marleen Onwezen *et al.*, 'Eating insects. How to make it the new normal', Wageningen Economic Research, 2018, E-depot WUR.
6   Marleen Onwezen *et al.*, 'Consumenten meer geneigd om "alternatieve" eiwitten te eten dan in 2015', Wageningen University & Research, 2020, E-depot WUR.
7   Marleen Onwezen *et al.*, 'Consumer acceptance of insects as food and feed: The relevance of affective factors', *Food Quality and Preference* 77 (2019) 51–63.
8   Frans Boogaard, 'Voor het eerst keurt Brussel insect goed voor consumptie: na meelworm volgen er zeker méér', *AD*, 4 May 2021.
9   Jonas House, 'Insects as food in the Netherlands: Production networks and the geographies of edibility', *Geoforum* 94 (2018) 82–93; Wageningen Universiteit, 'Waarom insecten niet de nieuwe sushi zijn', 5 October 2018.
10   Ingvar Svanberg & Åsa Berggren, 'Insects as past and future food in entomophobic Europe', *Food, Culture & Society* 24 (2021) 624–38.

CHAPTER 6

Milk, the Dutch white engine

Drinking milk: a special custom

Milk is good for everyone. The biblical land of milk and honey. And 'Milk, the white engine', the Dutch Dairy Agency's advertising campaign in the 1980s. Milk is a peculiar product: loved by us, but unknown, unloved and even loathed in other parts of the world. American anthropologist Marvin Harris even goes so far as to divide the world into lactophiles and lactophobes: the milk lovers and the milk haters.

Milk has many benefits as a food source: it is a rich source of energy, protein, vitamins A and D, and minerals such as calcium. Vitamin D is needed, along with calcium, for bone development. Humans can produce vitamin D themselves through sunlight, but at Europe's higher latitudes the sun shines less than in the Mediterranean. Any vitamin D deficiency can be supplemented by dairy consumption. This is made possible by a centuries-long evolutionary process involving frequent milk consumption, which developed a gene in the peoples of northern Europe ensuring that people can tolerate milk well beyond infancy.

A special drink

Milk is a special drink because it is a thirst quencher of animal origin. Most other thirst-quenching drinks consist of water with a vegetable component. Beer used to be the most reliable thirst quencher, but nowadays we can choose from tea, herbal tea, coffee, fruit drinks and the infinite range of soft drinks. I haven't include spirits and wine among thirst quenchers here, because they cannot provide the amount of water required for humans on a daily basis. This amount varies depending on a person's age and climate zone, but in the Netherlands, the *Voedingscentrum* recommends 1.5 to two litres of water a day for adults. Milk is also special because we do not need to slaughter the lactating animal for milk. Some of the milk intended for the young, lactating animal is used for human consumption.

Blood is edible

There is another custom, somewhat similar to milking animals: draining blood from cattle. The animal is then not killed. The blood obtained is a foodstuff. The custom occurs among the cattle-rearing peoples of East Africa: the Masai and the Turkana. They tap blood from the vein at the neck of (preferably) an ox. The blood is then drunk fresh, sometimes mixed with milk. The clotted blood can also be eaten.[1] So for the Masai and Turkana, blood is 'edible'.

© ADEL P. DEN HARTOG, 2024 | DOI:10.3920/9789004701571_008

By contrast, in other cultures, including those that follow Jewish and Muslim food laws, the use of blood as a foodstuff is actually forbidden.

Source

1   Michael Little, Sandra Gray, Benjamin Campbell, 'Milk consumption in African pastoral peoples'. In: Igor de Garine & Valerie de Garine (ed.). *Drinking. Anthropological approaches* (New York: Berghahn Books, 2001) 66–86, esp. 74–5; C. Daryll Forde, *Habitat, economy and society. A geographical introduction to ethnology* (London: Methuen, 1961) 295–6.

Milk in food culture

Drinking milk is inextricably linked to Dutch and other northern European food cultures, although this region is also home to people who dislike milk. In other parts of the world, such as China and Southeast Asia, milk has long remained relatively unknown as a food. On a global level, drinking milk is still an exceptional habit: some 70 percent of the world's population virtually never drinks milk again after breast milk. Incidentally, the classic dichotomy between milk drinkers and non-milk drinkers is changing. Eating habits are not stable. In today's China and Vietnam, milk and milk products are finding acceptance, although milk consumption is unlikely to reach as high a level as in the Netherlands for the time being.

The history of dairy consumption

Based on the 1896 study by German geographer Eduard Hahn (1856–1928) on the relationship between humans and domestic animals, cultural geographer Frederick Simoons made a classification of peoples who traditionally do and do not use milk. Tropical Africa, Southeast Asia and China, Simoons wrote, are areas without dairy farming traditions. Dairy cattle are virtually non-existent there, so milk is not part of conventional diets. Milk was also absent from the diet of the original inhabitants of the Americas and Australia. Cattle, goats and sheep were absent from these parts of the world in the past.

The Fertile Crescent

The use of milk probably originated around 7000 BCE in the Middle East, in what is called the Fertile Crescent. This area stretched from Israel-Palestine, Syria, southern Turkey to southern Iraq. It was a region rich in flora and fauna, where hunters and gatherers eventually started growing wild grasses (grains) and rearing animals. This probably happened first with goats and sheep around 11,000 years ago, followed by cattle and pigs. The moment people started milking their cattle is a very significant event in food history. As for milk consumption, it is likely that it started with milking goats and sheep. As milk spoils quickly without processing, these early herders and farmers will have quickly started using techniques to make milk more sustainable in the form of sour fermented milk, yoghurt, cheese and butter.

Milk in different cultures

From the Middle East, cattle breeders and farmers spread to Europe, North and East Africa, Central Asia and the Indian subcontinent. Dairy farmers, called the Linear Pottery Culture people after their typical pottery (Figure 42), reached the northern Balkans around 6500 BCE and spread further across Europe from there. Around 1,200 years later, they reached southern Limburg. They kept cattle not only for meat, but also for milk and dairy products such as cheese. According to several researchers, they laid the foundation for the emergence of the dairy farming so characteristic of northern Europe. Later peoples, such as the Funnel Beaker peoples (4300–2800/2700 BCE), who built the dolmens (in Dutch, *hunebedden*), were also cattle farmers. The Dolmen Centre in Borger shows that they kept cattle, sheep

FIGURE 42   Band ceramics from Steinheim an der Murr, *c.*5200 BCE. Collection
             Landesmuseum Württemberg
             PHOTO BY USER DADEROT, WIKI COMMONS

and goats. There is evidence that milk and milk products, probably from sheep and goats, were well known among these farmers. Cattle, on the other hand, were probably used more as draught animals.

The Middle Ages

Drinking milk, as we know it today, was not a common practice in the Netherlands in the Middle Ages. Nor was it seen as healthy. However, milk was widely used in poultices. The upper classes associated milk and dairy products with food customs of peasants and looked down on them. Characteristically, the revolt of hungry peasants in Kennemerland and West Friesland, 1491–1492, is known as the revolt of the cheese and bread people. The rebels carried banners on which these products were painted to show what they were fighting for (Figure 43). Dairy and bread formed the basic diet of the countryside, while the upper classes in the city mainly ate meat. This is also evident in the *Kerelslied* from the 14th century. This Flemish satirical song portrays peasants as stupid gluttons.[1] The refrain reads:

WOEDE van het KAAS en BROODS VOLK te ALKMAAR.

FIGURE 43
Anger of the Cheese and Bread people at Alkmaar (in 1492). In the background the cheese and bread flags
DRAWING BY JACOBUS BUYS, 1785, RIJKSMUSEUM COLLECTION

---

1   Paul Claes, *Lyricism of the Low Countries. The canon in eighty poems* (Amsterdam: De Bezige Bij, 2008) 70–71.

Wronglen wey broot ende caes Dat heit hi al den dach
Daer omme es de guy so daes Hi ites more than hys mach

Curd and whey, bread and cheese He eats that all day long
That's why the farmer is so lame He eats more of it than he can eat

Distribution

From the Fertile Crescent, the use of milk spread to North and East Africa, Central Asia and the Indian subcontinent. Milk still plays an important role in the diet of nomadic cultures in the arid regions of Africa. I wrote an article in *Voeding* (1980) about the role of dairy products among the Peul or Fulani, itinerant peoples of West Africa. They have all kinds of expressions highlighting the importance of milk, such as 'where there is milk, there is no hunger'. Illustrating the importance of milk, Africa correspondent Koert Lindijer described in 2011 how, after many years of drought in Kenya, the barren sandy plains turned green again and cattle gave milk again. The joy among the semi-nomadic Samburu was great: all the cows were back home and everyone was drinking milk again.

From India around the year 1000, the use of milk spread to Southeast Asia through the growing influence of Hinduism and later Buddhism. For ritual use, zebu cattle were kept around temples and other shrines as dairy cattle. However, it remained only for ritual use: the custom of drinking milk did not find acceptance among the population. After the arrival of Islam on Java in the 15th century, these ritual customs also disappeared.

The spread of milk to non-milk-drinking parts of the world gained new momentum with European expansion in the 16th and 17th centuries. European settlers took cattle, sheep and goats—their pets—to the Americas. In the eighteenth century, the English introduced these animals to Australia and New Zealand. In the VOC era, the seventeenth and eighteenth centuries, Dutchmen brought dairy cattle to the Dutch East Indies for their own use. In my thesis *Diffusion of milk as a new food to tropical regions: the case of Indonesia 1880–1942,* I wrote about several dairy farms that existed in Java in the first half of the 20th century. After Indonesia's independence, milk production continued to exist only on a modest scale in Java.

Milk doesn't just come from the cow

Unknown or lesser known types of milk in the Netherlands include donkey milk, buffalo milk, camel milk, mare milk, and yak milk. Buffalo milk is best known in India and Pakistan, but also in Italy. There, the buffalo was introduced in the eighteenth century, and its milk is processed into mozzarella, a soft cheese. The use of milk from camels and dromedaries is still limited to the dry and hot areas of North Africa and Asia, where these animals are very well adapted to the extreme conditions (Figure 44).

FIGURE 44   A Rabari man from India with camel milk
PHOTO FROM SAHAPEDIA

A special type of milk is mare's milk. Among the Turkish-speaking peoples of Central Asia, the Kazakhs, Kyrgyzs and Turkmens, *koemis*, fermented mare's milk, is very popular. Since mare's milk contains a higher percentage of lactose than cow's milk, it can be fermented to make this light-alcoholic drink. *Koemis* is a distinctive part of the food culture, even—despite the two percent alcohol—for the Muslim peoples of Central Asia.

The cosmetic use of donkey milk goes back as far as ancient Egyptian times. Queen Cleopatra is said to have bathed in donkey's milk to preserve her youth and beauty. From the late nineteenth century, donkey stables existed in the Netherlands to provide milk for infants. Donkey milk was said to be easier for infants to digest than cow's milk because its composition was more like that of human breast milk. In the 21st century, donkey dairies were again set up in our country.

Apart from donkey milk, horse or mare milk and horse milk products are sold in the Netherlands. They are touted as nutritional supplements. According to producers, they have beneficial effects for people with skin conditions and they improve health.

Explanations for the absence of milk in the diet

Why does milk barely feature in the diets of China, Southeast Asia, tropical Africa, and that of the indigenous peoples of Australia, New Zealand and Oceania? After all, we are talking about more than 2.5 billion people. Do they find milk abhorrent, are they allergic to it, or do they just not like the taste? Four interdependent factors—economic, geographical, genetic and cultural—play a role.

The economic factor

To start with the economic factor: where there is no significant dairy farming, milk is scarce and therefore unaffordable for most consumers. Even in a country like India with a centuries-old dairy tradition, milk is unaffordable for some 30 percent of the population living in abject poverty. For the wealthier groups, milk products are imported. Even in China, poor people cannot afford milk. However, interest in milk and dairy has increased among the emerging middle class. Besides milk powder for infants and young children, milk drinks in a variety of flavours are therefore sold in small cartons. So despite the fact that China does not have a typical milk culture, the country still represents a growth market for FrieslandCampina. As Friesche Vlag, this Dutch company was already selling condensed, sweetened milk in cans in Hong Kong and the former Dutch East Indies in the 1930s to Europeans living and working there. Condensed milk of various brands is still on sale in Asian countries (Figure 45).

FIGURE 45   A wide range of condensed milk in a shop in Hong Kong
PHOTO BY USER NAIMCBIM WOADD, WIKI COMMONS

The geographical factor

The main geographical factor is the problem of cattle rearing in the tropical parts of Africa and Southeast Asia. The high temperatures are unfavourable for cattle and milk yields. In tropical Africa, the blood-sucking tsetse fly is an additional obstacle. This fly causes the serious and ultimately deadly sleeping sickness among mammals. Cattle in these places could therefore never become part of the agricultural system and therefore not part of the food culture either. There were no cattle, sheep, goats or horses in America before the arrival of Europeans. The same is true of Australia before the arrival of the English in the eighteenth century. No cattle bred there, or in Oceania. In the Andes, the great indigenous civilisations and the Incas did domesticate the llama and alpaca as pack animals, for wool and, to a lesser extent, for food. However, these animals were not used as dairy cattle.

The geographical explanation is more complicated when we look at China and Japan. Climatically, it is quite possible to rear dairy cattle there, and cattle were not entirely unknown there. It seems that the spread of cattle rearing and the habit of drinking milk from Mesopotamia did not extend beyond Mongolia. However, during the Tang dynasty (618–907), a cultural exchange took place in the border areas of northern China between the Mongolian nomads with their dairy traditions and the people of northern China. Milk and dairy products found their way into the aristocracy of northern China through contacts with prominent nomadic families. This is a striking development as there was an aversion to the use of milk associated with Indians, Tibetans, and Central Asian nomads, the despised neighbouring peoples. The importance of milk and dairy then disappeared again in China with the decline of nomadic influences and the increasing scarcity of land, which was needed both for cattle rearing and food production.

The melamine scandal

The demand for milk powder for infant formula is so high in China these days that it is apparently worth messing with milk. In 2008, China was rocked by the *melamine* scandal. *Melamine* is a synthetic substance that was mainly used to produce plastics. Milk traders added this cheap substance to milk to increase its nitrogen content. The price of milk is largely determined by its protein content, for which the nitrogen content is a good measure. Milk powder with *melamine* was therefore more profitable, but the fraudsters overlooked the fact that *melamine* causes severe kidney diseases in babies and young children. In the end, six babies died and more than 300,000 children became ill.

Source
Paul Roberts, *The end of food. The coming crisis in the world food industry* (London: Bloomsbury, 2009).

The genetic factor

A major factor contributing to low milk consumption in some regions is the prevalence of genetically determined lactose intolerance. Milk—including mother's milk—contains milk sugar, known as lactose. Lactose is absorbed in babies in the small intestine after being broken down into glucose and galactose by the enzyme lactase. However, if this enzyme is absent or insufficient, lactose cannot be processed, causing nausea, abdominal cramps and diarrhoea. In the literature, the absence of the enzyme lactase is also called lactase deficiency. A minority of the world's population, especially descendants from north-western Europe, maintain high lactase activity levels throughout their lives and do not experience digestive problems with lactose (Figure 46).

In some 75–80 per cent of the world's population, lactase production disappears after two to five years, leaving these children and adults unable to tolerate large amounts of (cow) milk. Thus, their aversion to milk has a genetic basis. It is important to note here that breast milk contains more lactose (6.3–7%) than cow's milk (4.4–4.9%). Populations with lactose intolerance can generally drink milk in small quantities without any problems. Fermented milk products such as buttermilk, yoghurt and the aforementioned *koemis* also cause fewer problems because they contain lactase and thus partly help break down lactose.

Some people have a congenital defect that means they do not produce the enzyme lactase, or produce it insufficiently, even as babies. These people are therefore unable to consume milk. This is therefore different from lactose intolerance.

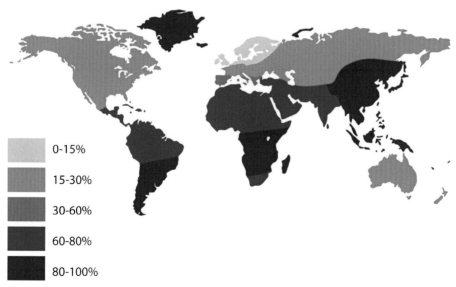

0-15%

15-30%

30-60%

60-80%

80-100%

FIGURE 46   Distribution of lactase deficiency around the world
BY FOOD INTOLERANCE NETWORK, WIKI COMMONS

There are people who are allergic to cow's milk. This too is not intolerance, although the symptoms of both conditions are sometimes similar.

The cultural factor

As discussed in Chapter 3, there is a particular prohibition in Jewish food laws on the use of milk: it must not be combined with meat. Less well known is that this prohibition also occurs among the pastoral peoples of East Africa. According to some researchers, this custom might have arisen in the past through contacts between Jewish traders and East African cattle breeders, but there is no concrete evidence for this. Most likely, therefore, this prohibition originated independently in East Africa. Among the pastoral peoples on the other side of the continent, in West Africa, the Ahaggar Tuaregs (in the extreme south of Algeria) are the only ones who also keep milk and meat separate. But even they don't know exactly how this prohibition arose.

The modern era

With the advent of modern dairy production after 1870, drinking milk became increasingly important. The dairy industry, supported by food science, has popularised milk as a necessary and indispensable healthy drink since the economic crisis of the 1930s. In Britain, *Milk Marketing Boards were* already active before World War II. In Scandinavian countries, too, the dairy industry and the government worked closely together to promote milk consumption.

Advertising campaigns

Although milk and dairy products were an established part of the Dutch food culture, the need to increase consumption arose in the 1930s. The means for this was advertising. On the initiative of the dairy organisations, the Crisis Dairy Bureau was set up in 1934. Cheese and other dairy products were advertised at exhibitions, 'dairy days' and cooking demonstrations. After World War II, these activities were taken over by the Dutch Dairy Agency, which was established in 1950. This agency's first major milk campaign began in 1958. With great success, the M-brigade was established, boasting half a million children as members. This campaign was followed in 1965 by the launch of Joris Driepinter, a little man featured in advertisements as being capable of all sorts of trials of strength because of drinking milk. Successive campaigns were sometimes a little forceful in their message ('I drink milk. You too?' and 'You need milk. Milk does you good.'), and achieved the opposite effect. Young people increasingly resisted milk. The 'Milk, the White Engine' campaign, from 1983 onwards, had no visible effect on milk consumption either. Finally, in 2003, the Dutch Dairy Agency's collective milk propaganda came to an end.

»MELK DE WITTE MOTOR«

NEDERLAND KRIJGT STARTPROBLEMEN

Loesje

Postbus 1045
6801 BA Arnhem
www.loesje.nl

FIGURE 47
Melk de witte motor. Loesje, 2 June 2008
REPRODUCED WITH PERMISSION FROM
LOESJE

'Milk, the White Engine' became a term that was even used in a poster by Loesje, an international free speech organisation (Figure 47), but the advertising campaigns could not turn the tide. Milk consumption continued to decline. We see similar trends in other western European countries. As a result of the increase in prosperity, eating patterns have shifted from relatively cheap to more expensive foods. As a result, milk had to compete with cola and other soft drinks, fruit drinks and mineral waters. In the Netherlands, milk consumption fell even further in the 21st century.[2]

Food aid

Objections to milk use also gained an international dimension in the 1980s when the European Community and the United States made powdered milk available for food aid. The starting point was simple: Western countries were struggling with dairy surpluses, so why could these not be used in the form of milk powder to combat protein shortages and malnutrition in developing countries? The main objections, however, were that milk did not fit into developing countries' diets and that milk powder might replace mother's milk. The latter objection was based on the notorious brochure *The baby killer* (1974), which outlined the fatal consequences for babies and young children of using milk powder mixed with contaminated water in unhygienic feeding bottles. The use of skimmed milk powder, which lacked vitamin

---

2   NOS, 'Nederlanders kopen steeds minder melk, vla en toetjes', 5 August 2020.

A and D, was also controversial. Indeed, deficiencies in vitamin A can lead to serious eye damage.

Criticism of this form of dairy food aid was justified in many ways, as its practical implementation left much to be desired. After initial confusion about what to do and what not to do, the Dutch dairy industry took responsibility by carefully handling food aid. This was done in close consultation with experts from the food world, in which I myself was involved. The starting point was to scrupulously consider whether every application for dairy aid was nutritionally and ethically sound.

Positive or negative effects?

The discussion about drinking milk received new impetus with the publication of the book *Eat, drink, and be healthy* (2001). This American bestseller on food and health was written by the controversial nutritionist Walter Willett (1945–). According to him, high dairy consumption increases the risk of developing prostate and breast cancer and cardiovascular disease. The widespread occurrence of lactose intolerance, according to Willett, is proof that milk is not suitable for humans. His conclusions are not always shared by other food scientists. Moreover, he ignores the food culture of milk-using peoples. In a later Wageningen-American study, in which Willett also participated, his standpoint was toned down.[3] Research into the effects of milk and milk products continues; we certainly haven't heard the last word on it.

A 1928 pamphlet on milk

In 1928 visual artist Erich Wichman (1890–1929) published an extraordinary pamphlet against milk consumption, entitled *The White Peril. On milk, milk use, milk abuse & milkiness*. Wichman was annoyed by the activities of the 'Amsterdamsch Drankweer Comité' against drinking. Milk saloons (Figure 48) were employed as an alternative to the pub. He considered milk the drink of mediocrity. Among other things, he wrote:

> For milk is not good for everyone, but milk is naturally good only for young calves [...] Milk abuse is one of the main causes of "modern" "nervousness" [...] certainly for an adult man milk is more harmful than gin [...] We are rotting in the milk pool. The white tide swallows us up.

---

3   Sabita Soedamah-Muthu *et al.*, 'Milk and dairy consumption and incidence of cardiovascular diseases and all-cause mortality: dose-response meta-analysis of prospective cohort studies', *The American Journal of Clinical Nutrition* 93:1 (2011) 158–71.

FIGURE 48
Theodora Johanna Hein's milk
parlour on the Bemuurde Weerd O.Z.
in Utrecht, ca. 1912–1915
PHOTOGRAPH BY E. SANDERS,
UTRECHT CITY ARCHIVE

However these discussions play out, it will not discourage milk drinkers. Milk and milk products have been used and appreciated by humans for some 8,000 years. Interestingly, after the second half of the 20th century, milk became more prevalent in the emerging middle classes of countries where milk was unknown. Hence, the saying 'unknown is unloved' is becoming less and less true as far as milk is concerned.

More about milk

Netherlands
In the Netherlands, figures from the Food Consumption Survey 2012–2016 (VCP) show that daily dairy consumption has decreased significantly, by 12 percent compared to the survey between 2007 and 2010. The average dairy consumption is 352 grams per day, of which 42 percent is milk, 15 percent yoghurt and nine percent cheese.[1] On average, a Dutch person eats about 20 kilos of cheese per year, which is above the European average.[2] In the period 2016–2019, the consumption of quark, yoghurt and cream increased, but all other dairy products were consumed less. Milk and buttermilk fell particularly sharply, by 6.1 and 7.8 percent respectively.[3]

Developing countries

While milk consumption is declining in the Netherlands, for vulnerable households in poor countries, the FAO actually aims to increase the use of milk and dairy products (Figure 49). For these people, dairy is an essential addition to a diet that continues to consists mainly of carbohydrates. Dairy contains the necessary proteins, fats and minerals missing from this unvaried diet. However, the price of dairy products hinders their consumption by the very people who need them most. As a solution, the FAO is trying to promote the consumption of milk from species other than cows. It is mainly goats and sheep that qualify for this, but the milk of buffalo, reindeer, moose, llamas, alpacas, horses, donkeys, yaks and camels is also excellent.[4]

Figure 50 shows growth rates of dairy consumption as expected by the Organisation for Economic Cooperation and Development (OECD) in developed and developing countries. In developing countries, they expect growth to lag behind. To increase consumption, investment in local production methods is needed above all. Improving the way dairy cattle are kept, fed and reared will increase production, the OECD says. Better transport facilities, cooling and processing methods will also have to contribute structurally to more products of better quality.

FIGURE 49   Photo on the cover of the FAO report: Milk and Dairy Products in
Human Nutrition (2013)
SOURCE: FAO OF THE UNITED NATIONS. REPRODUCED WITH
PERMISSION

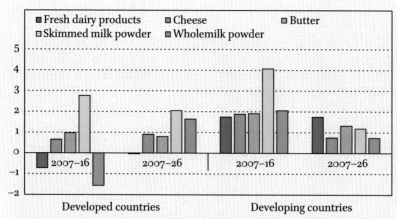

FIGURE 50   Growth rates of per capita consumption of dairy products
SOURCE: OECD-FAO AGRICULTURAL OUTLOOK 2017-2026

New economies

In recent decades, the biggest changes in dairy production and consumption have been seen in Asia. Although here too the differences between countries are considerable, dairy consumption is increasing everywhere. Between 1981 and 2007, per capita milk consumption doubled.[5] This is a striking development because lactose intolerance is common among the Asian population due to a genetically determined decrease in lactase activity. Most people who are effectively lactose intolerant can tolerate a certain amount of milk and milk products.

China stands apart in every respect. In China, dairy consumption is rising sharply with government support. For example, there is a school milk programme. Yoghurt in particular is becoming popular. The changing consumption pattern in China is closely linked to the increasing prosperity of the population. Greater purchasing power and openness to the West are tempting some Chinese to change traditions. For them, dairy is part of a 'Western lifestyle'.[6] Large dairy factories have been built in recent years (Figure 51), but now many more cows need to be reared to cater for production.

By 2022, China's dairy market is already expected to be larger than that of the U.S. Although dairy production in China is increasing, demand remains even higher. Therefore, Dutch dairy companies are spotting opportunities. COVID-19 did put some pressure on foreign trade,[7] but a China Dairy Exhibition was again organised in China in October 2020.[8] The Dutch dairy industry was present, as was the Sino-Dutch Dairy Development Centre, the research and knowledge centre established in 2013 by FrieslandCampina, Wageningen UR and China Agricultural University.[9]

Sources

1   Nederlandse Zuivel Organisatie, 'Daling zuivelconsumptie reden tot zorg', 21 November 2018.
2   ZuivelNL. 'Publicatie Zuivel in Cijfers 2019', 21 July 2020.
3   Erik Colenbrander, 'Yoghurt en kwark in trek, melkconsumptie omlaag – Nieuws Zuivelconsumptie', *Boeren Business*.
4   Ellen Muehlhoff, Anthony Bennett and Deirde McMahon, *Milk and Dairy Products in Human Nutrition*, Rome: FAO, 2013.
5   Nancy Morgan, 'Introduction: Dairy development in Asia', in: idem (red.), *Smallholder dairy development: Lessons learned in Asia*. Bangkok: FAO, 2009).
6   Daxue Consulting, 'The dairy market in China will be the world's largest by 2022', 1 June 2020.
7   Jacomien Voorhorst, 'Geopolitiek zorgenkindje voor FrieslandCampina', Nieuwe Oogst, 23 October 2020.
8   Ministerie van Landbouw, Natuur en Voedselkwaliteit, 'Report of the China Dairy Expo 2020', 13 October 2020, Agroberichten buitenland.
9   Ministerie van Landbouw, Natuur en Voedselkwaliteit, '49 keer op en neer naar China voor zuivelcentrum', 9 January 2020, Agroberichten buitenland.

Is a change as good as a feast?

In this book, Adel den Hartog has shown how complex people's food choices have been for millennia. If you ask why we eat what we eat, you end up with practical issues such as product availability, transport and cost, but also physiological, cultural, geographical, and psychological factors. Den Hartog points out the importance of the access individuals and groups have to food, and of their socio-economic position.

There is a special focus in this book on food culture. Of course people need to eat and drink to survive, but the meaning they have given to their consumption now and in the past is far from purely physiological. Religious beliefs, rituals, and other (ethical) views have led to a variety of food taboos. What is considered edible and inedible is therefore relative, depending on place, culture and time period.

Such dietary rules, shaped by practical factors as well as ideas about food, have historically had important symbolic value. Within societies, the rules often promoted social cohesion and formed part of the group identity. Thus, what people ate and drank reinforced the idea of an 'us', but also of a 'them': the other was a person who could not or would not follow the group's eating habits. Because people were, and often are, part of different cultural groups, individuals often move between multiple, overlapping food cultures.

Today, many different parties are trying to adjust people's eating habits. NGOs are pushing for a protein transition to reduce the far-reaching impact of our diet on the climate. Their call is linked to growing concerns about the sustainability of today's consumer society. Governments are trying to get citizens to choose foods that contain less fat, sugar and salt, hoping to make and keep them healthier. They fear that rising healthcare costs are an urgent reason for radically changing our diet. At the same time, multinationals are trying to increase the market for their products, some healthy, some not so healthy. In doing so, they use both the 'old-fashioned' printed press and viral marketing via social media.

Can someone be persuaded to eat something they regard as inedible? Adel den Hartog's work shows that stakeholders may need to adjust their ambitions. The idea that raw food is rabbit food and the religious belief that pork is unclean did not come out of the blue. They are parts of traditions. Sometimes—through a complex interplay of geographical, physiological, cultural and economic factors—groups have been excluding certain foods for centuries. Den Hartog gives many examples of our lack of flexibility, whether prompted by a genetic intolerance, or because what was unknown remained unloved.

Yet food habits can change. Sometimes food taboos disappear. Whereas wealthier citizens in the United States used to prefer not to eat the ubiquitous oyster because they saw it as 'poor man's food', this mollusc is now a standard part of the menu in fancy restaurants. Conversely, the increased availability of a crop or product can also increase its popularity in a culture, even if—as in the case of the tomato—it is sometimes slow. Ethical concerns can also become mainstream. Animal welfare, a concern only of a marginal group of people in the Netherlands at the beginning of the 20th century, is now higher on the agenda. Moreover, people are concerned about climate change. As a result of these factors, today 41 percent of Dutch people think that eating meat on a daily basis is a thing of the past.

Adel den Hartog shows that organisations committed to change can be cautiously optimistic. Certainly, food habits are slow to respond to interventions. Moreover, some social norms are far from malleable and some cultural practices almost irreversible. But people's views on what they see as 'edible' and 'inedible' can change. Without that flexibility, Homo sapiens would have died out long ago.

Annemarie de Knecht-van Eekelen and Jon Verriet
Malden, September 2021

About Adel den Hartog

Adel Peter den Hartog was born on 20 April 1937 in the city of Groningen. In the province of the same name, his father, Cornelis (Cees) den Hartog (1905–1993), had a general practice. At the beginning of World War II, Cees den Hartog was appointed director of the Information Bureau of the Food Council (the present-day *Voedingscentrum*) so the family moved to The Hague in 1941. After secondary school, Adel then completed a degree in social geography at Utrecht University in 1966.

Adel then worked as a Nutrition Project Officer at FAO in Accra in Ghana, and later at FAO headquarters in Rome. His many working trips for this organisation were aimed at mapping food habits and needs of societies in West Africa.

In 1977, Adel was appointed to the Department of Human Nutrition at Wageningen University of Agriculture. His father had laid the foundations for this department between 1955 and 1972, first as special professor and from 1969 as full-time professor of nutrition and food preparation. As a social geographer, Adel added new aspects to Wageningen food research and education. The way in which cultures in the past

Adel P. den Hartog (1937–2012). Own collection

and present gave and are giving meaning to food became particularly important in the curriculum and within research with Adel's arrival.

In Wageningen, Adel's work followed two tracks. On the one hand, he remained active in West Africa, where he helped set up large cooperation programmes in countries such as Benin and Burkina Faso. There, he focused particularly on maternal and child nutrition in the belief that this was vital for improving the health of the local population. In 1986, he obtained his PhD with a historical study on the use of milk in Indonesia between 1880 and 1942.

On the other hand, Adel developed a strong interest in (Dutch) food history. Already in his first Wageningen publication from 1977 on nutrition during famine, his historical interest comes to the fore. This culminated in a body of relevant articles, chapters and collections on the historical ideas and practices surrounding nutrition in the Netherlands and abroad. In particular, he presented his historical work at conferences and in volumes of the International Commission for Research into European Food History (ICREFH). Even after his retirement, he remained fascinated by this theme. Partly because of this, even when he was ill, he continued working on what would eventually become his last chapter.

Adel died on 26 September 2012 in Bennekom. He was 75 years old.

Literature consulted

Arens W. The Man-Eating Myth: Anthropology and anthropophagy. Oxford: Oxford University Press, 1979.

Barneveld D. De oude banketbakkerij. Bussum: van Dishoeck, 1968.

Behringer W. Kulturgeschichte des Klimas. Von der Eiszeit bis zur globalen Erwärmung. München: Verlag C. Beck, 2007.

Benkheïra MH. Tabou du porc et identité en Islam. In: ME Bruegel, B Laurioux (red.). Histoire et identités alimentaires en Europe. Parijs: Hachette, 2002, 37–51.

Beyens L. De graangodin. Het ontstaan van de landbouw. Amsterdam: Olympus, 2004.

Bie R van der, Hermans B, Pierik C, *et al.* Smakelijk weten. Trends in voeding en gezondheid. Den Haag/Heerlen: Centraal Bureau voor de Statistiek, 2012.

Bieleman J. Boeren in Nederland. Geschiedenis van de Nederlandse landbouw 1500–2000. Amsterdam: Boom, 2008.

Birlouez E. A la table des seigneurs, des moines et des paysans du Moyen Age. Rennes: Editions Ouest-France, 2009.

Booth AH. Small mammals of West Africa. Londen: Longmans, 1966.

Bossema W (bewerking). De witte illusie. Zuivel in Nederland, de EG en de Derde Wereld. Een onderzoek van het Centrum voor Ontwikkelingswerk Nederland (CON). Nijmegen: Stichting Derde Wereld Publikaties, 1988.

Bottéro J. La plus vieille cuisine du monde. Parijs: Audibert, 2002.

Boulestin B, *et al.* Mass cannibalism in the Linear Pottery Culture at Herxheim (Palatinate, Germany). Antiquity. 2009;83(322):968–82. Besproken in NRC/Handelsblad, 5/6-12-2009.

Bramah E, Bramah J. Die Kaffeemaschine. Die Kulturgeschichte der Kaffee Küche. Stuttgart: Parkland, 1995.

Briel JCT van, Hartog AP den. Leidraad zuivelvoedselhulp. Wageningen: Nederlands Instituut voor de voeding, 1984.

Brothwell D, Brothwell P. Food in Antiquity. A survey of the diet of early peoples. Londen: Thames and Hudson, 1969.

Brouwer ID. Nutritional impact of an increasing fuel wood shortage in rural households in developing countries. A literature overview. Progress Food and Nutrition Science. 1989;13:349–61.

Brouwer ID. Fuel and food. A hidden dimension in human nutrition. Proefschrift. Wageningen: Wageningen Universiteit, Afdeling Humane Voeding, 1994.

Brugmans IJ. Paardenkracht en mensenmacht: sociaal-economische geschiedenis van Nederland, 1795–1940. 's Gravenhage: Martinus Nijhoff, 1976.

Burke P. Cultural Hybridity. Cambridge: Polity Press, 2010.

Buishand T, Houwing HP, Jansen K. Groenten uit alle windstreken. Een geïllustreerde gids. Utrecht: Het Spectrum, 1986.

Buitelaar M. Vasten is alles met je mond. Enkele notities over vasten in Ramadan. In: M Buitelaar, GJ van Gelder (red.). Eet van de goede dingen. Culinaire culturen in het Midden-Oosten en de Islam. Bussum: Coutinho, 1995.

Burema L. De voeding van Nederland van de middeleeuwen tot de twintigste eeuw. Assen: van Gorcum, 1953.

Burnett J. Liquid pleasures. A social history of drinks in modern Britain. Londen: Routledge, 1999.

Burnett J. Plenty and want. A social history of diet in England from 1815 to the present day. Londen: Scolar Press, 1979.

Butler Flora C. Cattle. In: SH Katz, WW Weaver (red.). Encyclopedia of food and culture. New York: Charles Scribner's Sons, 2003. Deel 1, 330–40.

Castro BL. Food, morality, and politics: the spectacle of dog-eating Igorots and the 1904 St Louis World Fair. In: SR Friedland (red.). Food and morality. Blackawton: Prospect Books, 2008, 70–81.

Chang KC. Ancient China. In: KC Chang (red.). Food in Chinese culture. Anthropological and historical perspectives. New Haven: Yale University Press, 1977, 23–52.

Claes P. Lyriek van de Lage Landen. De canon in tachtig gedichten. Amsterdam: De Bezige Bij, 2008.

Comby B. Délicieux insectes. Les protéines du futur. Archamps: Editions Jouvence, 1989.

Dalby A. De keuken van Odysseus. Met 50 authentieke recepten. Amsterdam: Anthos, 2000.

Dam J van, Witteveen J. Koks en keukenmeiden. Amsterdamse kookboeken uit de Gastronomische Bibliotheek en de Bibliotheek van de Universiteit van Amsterdam. Amsterdam: Nijgh & van Ditmar, 2006.

Dantzig A van. Forts and castles of Ghana. Accra: Sedco Publishing, 1980.

Dantzig A van. Het Nederlandse aandeel in de slavenhandel. Bussum: Fibula van Dishoeck, 1968.

Darby WJ, Ghalioungui P, Grivetti L. Food: the gift of Osiris. Londen/New York/San Fransisco: Academic Press, 1977. Deel 1.

De Bijbel: De Nieuwe Bijbelvertaling uitgevoerd door het Nederlands Bijbelgenootschap. Amsterdam: Querido/Jongbloed, 2004.

De Koran. In vertaling van JH Kramers. Amsterdam: Agon Elsevier, 1974.

Delavigne AE. Pas de couchon, pas de Danois. Viande de porc et identité danoise, perspective anthropologique. In: ME Bruegel, B Laurioux (red.). Histoire et identités alimentaires en Europe. Parijs: Hachette, 2002, 37–51.

Doling A. Vietnam on a plate. A culinary journey. Hong Kong: Roundhouse Publications, 1966.

Dort-van Deursen AH van. Sint-Antonius en het varken. Brabantsheem. 1979;31(1):7–9.

Douglas M. Reinheid en gevaar. Utrecht/Antwerpen: Het Spectrum, 1976.

Drouard A. Horsemeat in France: A food item that appeared during the war of 1870 and disappeared after the Second World War. In: I Zweiniger-Bargielowska, R Duffett, A Drouard (red.). Food and war in twentieth century Europe. Farnham: Ashgate, 2011; Londen/New York: Routledge, 2016, 233–45.

Dufour DL, Sander JB. Insects. In: KF Kiple, KC Ornelas (red.). The Cambridge history of food. Cambridge: Cambridge University Press, 2000. Deel I, 546–54.

Ebbinge Wubben JC. Thema thee. De geschiedenis van thee en het theegebruik in Nederland. Rotterdam: Museum Boymans-van Beuningen, 1978.

Erasmus D. Het verbod vlees te eten en andere gelijkaardige bepalingen die van mensen uitgaan in een vertaling van John Piolon. Rotterdam: Donker, 2006/1522.

Faas P. Rond de tafel der Romeinen. Amsterdam: Rainbow, 1994.

FAO. Le lait et les produits laitiers dans la nutrition humaine. Rome: FAO, 1995.

FAO. The state of food and agriculture 2009. Livestock in the balance. Rome: FAO, 2009.

FAO. The state of food insecurity in the world 2010. Rome: FAO, 2010.

Fenton A. Milk and milk products in Scotland: the role of the Milk Marketing Boards. In: AP den Hartog (red.). Food technology, science and marketing: European diet in the twentieth century. East Lothian: Tuckwell Press, 1995, 89–102.

Ferguson G. Signs and symbols in Christian art. Oxford: Oxford University Press. 1976.

Forde CD. Habitat, economy and society. A geographical introduction to ethnology. Londen: Methuen, 1961.

Fouquet P, Borde M de. Histoire de l'Alcool. Que sais-je? nr. 2521. Parijs: Presse Universitaire de France, 1990.

Freeman M. Sung. In: KC Chang (red.). Food in Chinese culture. Anthropological and historical perspectives. New Haven: Yale University Press, 1977.

Fumey G, Etcheverria O. Atlas mondial des cuisines et gastronomie. Une geographie gourmande. Parijs: Éditions Autrement, 2004.

Fumey G. Approches géoculturelles de l'alimentation. In: Morinaux V (red.). Nourrir les hommes. Questions de géographie. Parijs: Editions du Temps, 2008, 27–37.

Gade DW. Horsemeat as human food in France. Ecology of Food and Nutrition. 1976;5:1–11.

Garine I de. The socio-cultural aspects of nutrition. Ecology of Food and Nutrition. 1972;1(2):143–63.

Geerts A. Wereldwijd is lactase-persistentie niet de norm. Voedings Magazine. 2011; 24(1):12–3.

Gísladóttir H. Horse. In: SH Katz, WW Weaver (red.). Encyclopedia of food and culture, 2003. Deel 2, 209–10.

Gloyer G. Albania. Chalfront St Peter: Bradt Travel Guides, 2008.

Grivetti LE. Dietary separation of meat and milk. A cultural-geographical inquiry. Ecology of Food and Nutrition. 1980;9:203–17.

Grivetti LE. Food prejudices and taboos. In: KF Kiple, KC Ornelas (red.). Cambridge world history of food. Cambridge: Cambridge University Press, 2000. Deel 2, 1495–513.

Goody J. Cooking, cuisine and class: a study in comparative sociology. Cambridge: Cambridge University Press, 1982.

Goudsblom J. Vuur en beschaving. Amsterdam: Olympus Amstel Uitgevers, 2009 4e druk.

Grooth METh de, Verwers GJ. Op goede gronden. De eerste boeren in Noord-West Europa. Leiden: Rijksmuseum van Oudheden, 1984.

Hahn E. Die Haustiere und ihre Beziehungen zur Wirtschaft des Menschen, eine geografische Studie. Leipzig: Duncker & Humblot, 1896.

Harris M. Good to eat. Riddles of food and culture. New York: Simon & Schuster, 1985.

Harris M. The cultural ecology of India's sacred cattle. Current Anthropology. 1966;7(1):51–4;55–66.

Hartog AP den, Staveren WA van, Brouwer ID. Food habits and consumption in developing countries. Manual for field studies. Wageningen: Wageningen Academic Publishers, 2006.

Hartog AP den. The changing place of vegetables in Dutch food culture: the role of marketing and nutritional sciences 1850–1990. Food & History. 2004;2(2):87–103.

Hartog AP den. Denk aan de vitamientjes. Groente in de Nederlandse eetcultuur. Spiegel Historiael. 2003;38(3–4):128–33.

Hartog AP den. Acceptance of milk products in Southeast Asia. The case of Indonesia as a traditional non-dairy region. In: K Cwiertka, B Walraven (red.). Asian food. The global and the local. Richmond, Surrey: Curzon Press, 2002, 34–45.

Hartog AP den. Constante en veranderende elementen van de Nederlandse eetcultuur: groente tussen traditie en trends. In: AP den Hartog (red.). De voeding van Nederland in de twintigste eeuw. Wageningen: Wageningen Pers, 2001, 109–19.

Hartog AP den. Changing perceptions on milk as a drink in Western Europe. The case of the Netherlands. In: I de Garine, V de Garine (red.). Drinking. Anthropological approaches. New York: Berghahn Books, 2001, 96–107.

Hartog C den, Hautvast JGAJ, Hartog AP den, Deurenberg P. Nieuwe Voedingsleer. Utrecht: Het Spectrum, 1988.

Hartog AP den. Diffusion of milk as a new food to tropical regions: the case of Indonesia 1880–1942. Proefschrift. Wageningen: Landbouwuniversiteit, 1986.

Hartog AP den. Veranderende voedselpatronen in ontwikkelingslanden en de rol van zuivelproducten. Voeding. 1980;41:292–302.

Hartog AP den, Vos A de. Knaagdieren (Rodentia) als voedsel in tropisch Afrika. Voeding. 1974;35:232–40.

Hartog C den. Kijk, voeding. Over mensen, hun voedsel en hun gezondheid. Utrecht: Het Spectrum, 1982.

Hendriks H, Verhagen H, Büchner F, et al. Gezondheidseffecten van groente en fruit. Voeding NU. 2010;12(3):26–8.

Hermus RJJ. De voeding van de schoolgaande jeugd: gezonde voeding of ongezondheidsopvoeding. Melk in relatie tot de Gezondheid. 1977;4(3):35–62.

Heston A. An approach to the sacred cow of India. Current Anthropology. 1971;12(2):191–209.

Hoek A. Will novel protein foods beat meat? Consumer acceptance of meat substitutes—a multidisciplinary research approach. Proefschrift. Wageningen: Wageningen Universiteit, Afdeling Humane Voeding, 2010.

Hofstee EW. De demografische ontwikkeling van Nederland in de eerste helft van de 19e eeuw. Een historisch-demografische en sociologische studie. Deventer: Van Loghum Slaterus, 1978.

Holas B. De nouveaux cas de cynophagie en Afrique Occidentale. Notes Africaines. 1955;nr. 65:22–3.

Hollander R den. Historische en actuele aspecten van de islamitische spijswetgeving: het principe van de islamitische spijswetten. In: M Buitelaar, GJ van Gelder (red.). Eet van de goede dingen. Culinaire culturen in het Midden-Oosten en de Islam. Bussum: Coutinho, 1995.

Holt VM. Why not eat insects? London: British Museum National History, 1885, reprint 1988.

Hortulus. De kloostertuin van Walafried Strabo. Vertaling V. Hunink. Toelichting Kruidentuincommissie Nederlands Openluchtmuseum. Warnsveld: Terra Lannoo, 2004.

Huis A van. Insecten als voedsel. In: T Huigens, P de Jong, *et al.* (red.). Muggenzifters en mierenneukers. Insecten onder de loep genomen. Wageningen: Laboratorium voor Entomologie, 2006, 247–55.

Isaac E. On the domestication of cattle. In: S Struever (red.). Prehistoric agriculture. Garden City, NY: The National History Press, 1971, 451–70.

Israel J. The Dutch Republic. Its rise, greatness, and fall 1477–1806. Oxford: Clarendon Press, 1995.

Itan Y, Powell A, Beaumont MA, *et al.* (2009). The origin of lactase persistence in Europe. Plos Computational Biology. 2009;5(8):1–13.

Jas J, Scharloo M, Wetering H van de. Het peperhuis te Enkhuizen. Zwolle: Waanders, 1996.

Jacobs EM. Koopman in Azië. De handel van de Verenigde Oost-Indische Compagnie tijdens de 18ᵉ eeuw. Zutphen: Walburg Pers, 2000.

Jobse-van Putten J. Eenvoudig maar voedzaam. Cultuurgeschiedenis van de dagelijkse maaltijd. Nijmegen: SUN, 1995.

Jong JJP de. De waaier van het fortuin. De Nederlanders in Azië en de Indonesische archipel 1595–1950. Den Haag: Sdu Uitgevers, 1998.

Jong L de. Vleesvervangers. Nieuwe generatie moet nog meer op vlees lijken. Voedingsmiddelen Industrie. 2010;11(1):26–7.

Kaplan P. Food in middle-class Madras households from the 1970s to the 1990s. In: K. Cwertka, B. Walraven (red.). Asian food. The global and the local. Richmund Surrey: Curzon Press, 2002.

Katan M. Wat is nu gezond? Fabels en feiten over voeding. Amsterdam: Bert Bakker, 2008.

Kemmers WH. De groente- en fruitveilingen tot 1945. In: P Plantenberg. 100 jaar veilingen in de tuinbouw. 's Gravenhage: Centraal Bureau van de Tuinbouwveilingen in Nederland en de Vereniging van Bloemenveilingen in Nederland, 1987, 12–34.

Kiado C. Gundel's Hungarian cookbook. Békéscsaba: Kner Printinghouse, 1986.

Kisbán E. Coffee in Hungary: its advent and integration into the hierarchy of meals. In: DU Ball. Kaffee im Spiegel europäischer Trinksitten. Zürich: Veröffentlichungen des Johann Jacobs Museums zur Kulturgeschichte des Kaffees, 1991. Deel 2.

Kisbán E. Goulash: a popular food item that became a national symbol. Boedapest: Életmód & Tradíció 4, 1989.

Kjaernes U. Milk: nutritional science and agricultural development in Norway, 1890–1990. In: AP den Hartog (red.). Food technology, science and marketing: European diet in the twentieth century. East Lothian: Tuckwell Press, 1995, 89–102.

Knecht-van Eekelen A de. Voeding en religie: over Joodse voeding in Nederland. In: A de Knecht-van Eekelen, M Stasse-Wolthuis (red.). Voeding in onze samenleving in cultuurhistorisch perspectief. Alphen aan den Rijn/Brussel: Samsom Stafleu, 1987, 93–109. Koningsbruggen W. Achtergronden en praktische tips bij lactose intolerantie. Voedings Magazine. 2010;23(3):12–5.

Koolmees PA. Symbolen van openbare hygiëne. Gemeentelijke slachthuizen in Nederland 1795–1940. Proefschrift. Rotterdam: Erasmus Publishing, 1997.

Lambert Oritz E. South America. In: SH Katz, WW Weaver (red.). Encyclopedia of food and culture. New York: Charles Scribner's Sons, 2003. Deel 3, 303–9.

Leach E. Claude Lévi-Strauss. Londen: Fontana/Collins, 1970.

Lévi-Strauss C. Le cru et le cuit. Parijs: Plon, 1964.

Leopold AC, Ardrey R. Toxic substances in plants and the food habits of man. Science. 1972;176(4034):512–4.

Leroi-Gourhan A. Milieu et techniques. Parijs: Editions Albin Michel, 1973.

Lignon-Darmaillac S. Les vignobles méditerranéens, In: V Moriniaux (red.). Questions de géographie la Méditerranée. Parijs: Edition de Temps, 2001, 161–72.

Little MA, Gray SJ, Campbell BC. Milk consumption in African pastoral peoples. In: I de Garine, V de Garine (red.). Drinking. Anthropological approaches. New York: Berghahn Books, 2001, 66–86.

Looker D. Pig. In: SH Katz, WW Weaver (red.). Encyclopedia of food and culture. New York: Charles Scribner's Sons, 2003. Deel 3, 74–81.

Lummel P. Food provision in the German army of the First World War. In: I Zweiniger-Bargielowska, R Duffett, A Drouard (red.). Food and war in twentieth century Europe. Farnham: Ashgate, 2011; Londen/New York: Routledge, 13–25.

Lyngo IJ. Symbols in the rhetoric on diet and health—Norway 1930: the relation between science and the performance of daily chores. In: A Fenton (red.). Order and disorder: the health implications of eating and drinking in the nineteenth and twentieth centuries. East Lothian: Tuckwell Press, 2000, 155–69.

Malaguzzi S. Eten en drinken. Gent: Ludion, Kunst Bibliotheek, 2007.

Malinowsky B. A scientific theory of culture and other essays. Chapel Hill, NC: University of North Carolina Press, 1944.

Matthaiou A. Prier comme un Turc et manger comme un chrétien, In: ME Bruegel, B Laurioux (red.). Histoire et identités alimentaires en Europe. Parijs: Hachette, 2002, 37–51.

Mauro F. Histoire du café. Parijs: Éditions Desjonquères, 1991.

McCollum EV. A history of nutrition. The sequence of ideas in nutrition investigations. Boston, MA: Houghton Mifflin Company, 1957.

Melk de nationale toverdrank. De Kleine Aarde. 1984;nr. 49.

Mérat MC. Dans l'intimité des Lutéciens. In: Paris raconte Lutèce. Les Cahiers Science & Vie. 2009;nr. 111,72–7.

Meulenberg MTG. Een eeuw verbruiksontwikkeling van tuinbouwprodukten. In: P Plantenberg. 100 jaar veilingen in de tuinbouw. 's Gravenhage: Centraal Bureau van de

Tuinbouwveilingen in Nederland en de Vereniging van Bloemenveilingen in Nederland, 1987, 169–93.

Mintz SW. Sweetness and power. The place of sugar in modern history. Harmondsworth: Penguin Books, 1985.

Moriniaux V. Les religions et l'alimentation. In: V. Morinaux (red.). Nourrir les homes. Questions de géographie. Parijs: Editions du Temps, 2008, 39–67.

Mulder GJ. De voeding in Nederland in verband tot den volksgeest. Rotterdam: HA Kramers, 1847.

Murton B. Australia and New Zealand. In: SH Katz, WW Weaver (red.). Encyclopedia of food and culture. New York: Charles Scribner's Sons, 2002. Deel 2, 1339–50.

National Geographic. State of the Earth 2010. Washington: National Geographic, 2009.

Navder KP, Lieber CS. Alcohol. In: SH Katz, WW Weaver (red.). Encyclopedia of food and culture, New York: Scribner, 2002/3. Deel 1, 62–6.

Niehof A. Food, diversity, vulnerability and social change. Research findings from insular Southeast Asia. Wageningen: Wageningen Academic Publishers, 2010.

Nijboer H, Tilburg B van. Tussen Compagnie en Handelsmaatschappij. De Nederlandse koffiehandel in de 18ᵉ eeuw. In: P Reinders, Th Wijsenbeek (red.). Koffie in Nederland. Vier eeuwen cultuurgeschiedenis. Zutphen: Walburg Pers, Gemeente Musea Delft, 1994.

Nijhof P. Emaille reclameborden in Nederland. Amsterdam: Van Soeren & Co, 1986.

Olsen SJ. Dogs, In: KF Kiple, KC Ornelas (red.). The Cambridge history of food. Cambridge: Cambridge University Press, 2000. Deel I, 508–16.

Panoff M, Perrin M. Dictionaire de ethnologie. Parijs: Payot, 1973.

Patou-Mathis M. Mangeurs de viande. De la préhistoire à nos jours. Parijs: Perrin, 2009.

Payne WJA. Cattle production in the tropics. Londen: Longman, 1970.

Pimentel D, Pimentel M. Food, energy and society. Londen: Arnold, 1979.

Plinius. De Wereld. Naturalis historia. Vertaald door J van Gelder, M Nieuwenhuis, T Peters. Amsterdam: Athenaeum-Polak & van Gennep, 2004.

Prins E. Leger en lichter. Genoegdossier vasten. 2010; nr 78:32–7, Genoeg.

Proef. Keurslagersmagazine. De Oostenrijkse keuken. Een heerlijke mengelmoes. 2010; no. 1.

Pyke M. Food and society. Londen: John Murray, 1968.

Roberts P. The end of food: The coming crisis in the world food industry. Londen: Bloomsbury, 2008.

Safran Foer J. Dieren eten. Amsterdam: Ambo/Anthos, 2010.

Salaman RN. The history and the social influence of the potato. Cambridge: Cambridge University Press, 1949.

Schafer EH. T'ang. In: KC Chang (red.). Food in Chinese culture. New Haven: Yale University Press, 1977, 87–140.

Schildermans J, Sels H, Willebrands M. Lieve schat wat vind je lekker? Het Koocboec van Antonius Magirus (1612) en de Italiaanse keuken van de Renaissance. Leuven: Davidsfonds, 2007.

Schlosser E. Fast food nation. The dark side of the all-American meal. New York: Harper, 2001.

Schreurs W. Collectieve reclame in Nederland. Leiden: Stenfert Kroese, 1991.

Scrimshaw NS, Murray EB. The acceptability of milk and milk products in populations with high prevalence of lactose intolerance. American Journal of Clinical Nutrition. 1988;48:1083–159.

Shaxson A, Dickson P, Walker J. The Malawi Cookbook. Zomba, Malawi: Government Printer, 1979.

Shephard S. Pickled, potted and canned. How the preservation of food changed civilisation. Londen: Headline Book Publishing, 2000.

Simoons FJ. Eat not this flesh: Food avoidances from prehistory to present. Madison: University of Wisconsin Press, 1994.

Simoons FJ. Questions of the sacred cow-controversy. Current Anthropology. 1979;20: 467–93.

Slicher van Bath B. De agrarische geschiedenis van West-Europa 500–1850. Utrecht: Het Spectrum, 1987.

Soedamah-Methu SS, Ding EL, Al-Delaimy WK, *et al.* Milk and dairy consumption and incidence of cardiovascular diseases and all-cause mortality: dose-response meta-analysis of prospective cohort studies. The American Journal of Clinical Nutrition. 2011;93:158–71.

Standage T. A history of the world in 6 glasses. Londen: Atlantic Books, 2007.

Takken W. Beten en steken. Hinderlijke insecten en andere plaaggeesten en hun effecten op onze gezondheid. Baarn: Trion Natuur, 2007.

Tannahill R. Food in history. St Albans: Paladin Frogmore, 1975.

Tannahill R. Vlees en bloed. De geschiedenis van het kannibalisme. Amsterdam: Wetenschappelijke Uitgeverij, 1975.

Taylor Sen C. Jainism: the world's most ethical religion. In: SR Friedland (red.). Food and morality. Blackawton: Prospect Books, 2008, 230–40.

Terlouw F. De aardappelziekte in Nederland in 1845 en volgende jaren. Economisch en sociaal-historisch jaarboek. 1971;34:264–308.

Tjon Sie Fat L. De tuin van Clusius. Leiden: Hortus Botanicus Leiden, 1992.

Topik S. Coffee. In: KF Kiple, KC Ornelas (red.). The Cambridge history of food. Cambridge: Cambridge University Press, 2000. Deel 1, 641–53.

Uytven R van. Geschiedenis van de dorst. Twintig eeuwen drinken in de lage landen. Leuven: Davidsfonds, 2007.

Verdonk DJ. Het dierloze gerecht: een vegetarische geschiedenis van Nederland. Amsterdam: Boom, 2009.

Volkers K. Wandelen over de bierkaai. Bierhistorische gids van Utrecht. Zaltbommel: Aprilis, 2006.

Waard JMD de. Voedselvoorschriften in Boeteboeken. Motieven voor het hanteren van voedselvoorschriften in vroeg-middeleeuwse Ierse boeteboeken 500–1100. Rotterdam: Erasmus Publishing, 1996.

Wallace AR (1869). The Malay archipelago. Singapore: Oxford University Press, 1869. New York: Harper & Brothers Publishers, reprint 1989.

Walraven B. Bardot soup and Confucian's meat. In: K Cwiertka, B Walraven (red.). Asian food. The global and the local. Richmond, Surrey: Curzon, 2002, 95–115.

War on Want. The baby killer. Londen: War on Want, 1974.

Weisberger JH, Comer J. Tea. In: KF Kiple, KC Ornelas (red.). The Cambridge history of food. Cambridge: Cambridge University Press, 2000. Deel 1, 712–20.

Whyte RO. Rural nutrition in monsoon Asia. Kuala Lumpur: Oxford University Press, 1974.

Wichman E. Het witte gevaar. Over melk, melkgebruik, melkmisbruik & melkzucht. Amsterdam: CJ Aarts, 2ᵉ druk. Amsterdamse Schotschriften 6, 1928/1979.

Wijsenbeek Th. Ernst en luim. Koffiehuizen tijdens de Republiek. In: P Reinders, Th Wijsenbeek (red.). Koffie in Nederland. Vier eeuwen cultuurgeschiedenis. Zutphen: Walburg Pers, Gemeente Musea Delft, 1994, 35–54.

Wijsenbeek Th. Grand-Café's en Volkskoffiehuizen. In: P Reinders, Th Wijsenbeek (red.). Koffie in Nederland. Vier eeuwen cultuurgeschiedenis. Zutphen: Walburg Pers, Gemeente Musea Delft, 1994, 127–48.

Willet W. Eat drink and be healthy. New York; Simon and Schuster, 2005.

Winock M. 1870: Paris assiégé par les Pruissiens. Les Collection de l'Histoire. 2000; nr. 9 Oct/Dec: 70–5.

Winter JM van. Spices and comfits. Collected papers on medieval food. Blackawton: Prospect Books, 2007.

Wirtz A. Die Moral auf dem Teller. Zürich: Chronos Verlag, 1993.

Wrangham R. Koken. Over de oorsprong van de mens. Amsterdam: Nieuw Amsterdam, 2009.

Wyatt AR. Mexico and Central America, Pre-Columbian. In: SH Katz, WW Weaver (red.). Encyclopedia of food and culture. New York: Charles Scribner's Sons, 2003. Deel 2, 497–502.